Laboratory Exercises in Organismal and Molecular Microbiology

Laboratory Exercises in Organismal and Molecular Microbiology

STEVE K. ALEXANDER Ph.D.
University of Mary Hardin–Baylor

DENNIS STRETE Ph.D.
McLennan Community College

MARY JANE NILES Ph.D.
University of San Francisco

 **Higher Education**

Boston   Burr Ridge, IL   Dubuque, IA   Madison, WI   New York   San Francisco   St. Louis
Bangkok   Bogotá   Caracas   Kuala Lumpur   Lisbon   London   Madrid   Mexico City
Milan   Montreal   New Delhi   Santiago   Seoul   Singapore   Sydney   Taipei   Toronto

Higher Education

LABORATORY EXERCISES IN ORGANISMAL AND MOLECULAR MICROBIOLOGY

Published by McGraw-Hill, a business unit of The McGraw-Hill Companies, Inc., 1221 Avenue of the Americas, New York, NY 10020. Copyright © 2004 by The McGraw-Hill Companies, Inc. All rights reserved. No part of this publication may be reproduced or distributed in any form or by any means, or stored in a database or retrieval system, without the prior written consent of The McGraw-Hill Companies, Inc., including, but not limited to, in any network or other electronic storage or transmission, or broadcast for distance learning.

Some ancillaries, including electronic and print components, may not be available to customers outside the United States.

This book is printed on recycled, acid-free paper containing 10% postconsumer waste.

1 2 3 4 5 6 7 8 9 0 QPD/QPD 0 9 8 7 6 5 4 3

ISBN 0–07–248744–5

Publisher: *Martin J. Lange*
Senior sponsoring editor: *Patrick E. Reidy*
Senior developmental editor: *Jean Sims Fornango*
Marketing manager: *Tami Petsche*
Senior project manager: *Gloria G. Schiesl*
Lead production supervisor: *Sandy Ludovissy*
Lead media project manager: *Audrey A. Reiter*
Senior media technology producer: *Barbara R. Block*
Designer: *K. Wayne Harms*
Cover/interior designer: *Scan Communications Group Inc.*
Cover images supplied by: *Dennis Strete*
Lead photo research coordinator: *Carrie K. Burger*
Photo research: *David Tietz*
Compositor: *Electronic Publishing Services Inc., NYC*
Typeface: *11/13 Times Roman*
Printer: *Quebecor World Dubuque, IA*

The credits section for this book begins on page 353 and is considered an extension of the copyright page.

Library of Congress Cataloging-in-Publication Data

Alexander, Steve K.
    Laboratory exercises in organismal and molecular microbiology / Steve K. Alexander,
Dennis Strete, Mary Jane Niles. – 1st ed.
        p. cm.
    ISBN 0–07–248744–5
    1. Microbiology—Laboratory manuals. 2. Molecular microbiology—Laboratory manuals.
I. Strete, Dennis. II. Niles, Mary Jane. III. Title.

QR63.A447   2004
579'.078—dc21   2002044883
                                                          CIP

www.mhhe.com

Dedication

To my family: Pamela, Steve Jr., Melissa, and Paul; and to my microbiology students: past, present, and future.

—SKA

To my family for their support and encouragement.

—DS

To my husband: John Higgins.

—MJN

Contents

S E C T I O N   V I
Controlling the Risk and Spread of Bacterial Infections   199

S E C T I O N   V I I
Bacterial Genetics   223

S E C T I O N   V I I I
Viruses   267

S E C T I O N   I X
Hematology and Serology   297

Preface

When students move from the lecture hall to the micro-biology laboratory, they need help bridging the gap between the theory and the practice of what they are learning. The equipment is unfamiliar, the procedures are unfamiliar, and many of the materials they are handling are unfamiliar. Linking the information from their classroom lectures to the laboratory procedures is necessary for their ultimate success. Our goal for this laboratory manual is to provide the bridge that helps students integrate their classroom lectures with their laboratory experiences. This integrated approach is the only way to ensure understanding and mastery in microbiology.

Features

- *Class-tested experiments* have been vetted in our own courses and provide a thoughtful progression of opportunities—from basic lab techniques, such as Exercises 9–15 on various staining techniques, to more challenging exercises, such as the simulated epidemic in Exercise 44: "Enzyme-linked Immunosorbent Assay (ELISA)." This building-block approach allows students to develop comfort and confidence in their laboratory skills.

- *Exceptional full-color art program* includes over 250 of our own photographs created specifically for these laboratory exercises, plus 150 line drawings of equipment, procedures, and results. Students can easily confirm their results and procedures by referring to the illustrations.

- *Exceptional attention to safety issues* is given throughout the manual. A basic lab safety section beginning on page xi includes a table identifying the biosafety level of every organism used in the experiments. The BSL 2 icon 🔵 appears where appropriate to remind students of the needed safety precautions when working with pathogens. Caution symbols ▼ appear throughout the lab manual to provide students with additional safety warnings as needed.

Organization

Our 46 exercises are organized into the following nine sections:

Section I	Survey of Microscopic Organisms
Section II	Staining Techniques
Section III	Bacterial Cultivation
Section IV	Bacterial Identification
Section V	Medical Microbiology
Section VI	Controlling the Risk and Spread of Bacterial Infections
Section VII	Bacterial Genetics
Section VIII	Viruses
Section IX	Hematology and Serology

The standard presentation of each section makes it easy for both students and lab managers to prepare for an exercise. Each exercise:

1. Opens with a short background that conveys only information relevant to the exercise.

2. Lists all needed materials, by category.

3. Presents procedures for the exercise in easy-to-follow steps and includes special notes, hints, and instructions to ensure success.

4. Integrates all photographs and line drawings into the text of the exercise where they will provide the student with the most support.

5. Includes a tear-out laboratory report conveniently located at the end of the exercise.

Instructor Support Material

An **Instructor Image Bank** provides digital files in the easy-to-use JPEG format for all of the photos and line art included in this lab manual. They are organized by section and placed in PowerPoint sets for easy access. These may prove useful for lab preparation packets, testing, or discussion sessions. Ask your McGraw-Hill representative for further details.

The **Instructor's Manual** for this set of laboratory exercises may be found online at:

www.mhhe.com/biosci/ap/labcentral/

It provides answers to lab report questions, tips for lab exercise success, and other useful information.

Acknowledgments

In the end, our hope is that we have put together a manual that will serve as a valuable teaching tool for the microbiology laboratory. Our efforts were greatly aided by the following reviewers, whom we gratefully acknowledge:

Daniel R. Brown, *Sante Fe Community College*
Kathy Buhrer, *Tidewater Community College*
Linda E. Fisher, *University of Michigan, Dearborn*
Georgia Ineichen, *Hinds Community College*
Hubert Ling, *County College of Morris*
Rita Moyes, *Texas A&M University*
Richard C. Renner, *Laredo Community College*
Ken Slater, *Utah Valley State College*

Kristin M. Snow, *Fox Valley Technical College*
Carole Rehkugler, *Cornell University*
Paul E. Wanda, *Southern Illinois University, Edwardsville*

Our gratitude is also extended to our publishing team at McGraw-Hill:

Colin Wheatley, Publisher/Sponsoring Editor
Jean Sims Fornango, Senior Developmental Editor
Tami Petsche, Marketing Manager
Gloria Schiesl, Project Manager
Sandy Ludovissy, Production Supervisor
Wayne Harms, Designer
Carrie Burger, Photo Editor

Safety Guidelines for the Microbiology Laboratory

General Guidelines for Every Lab Session

1. Wear appropriate clothing and shoes to the laboratory. Shoes must completely cover the feet to provide protection from broken glass and spills.
2. Place all books, backpacks, purses, etc., in an area designated by your laboratory instructor. Carry to your work area only the items you will use in the lab.
3. Wash your hands thoroughly with antibacterial soap before beginning the lab session.
4. Wipe your work area with disinfectant, and allow to air-dry before beginning the lab session.
5. Do not perform activities in the lab until you are given instructions by your laboratory instructor.
6. Do not eat, drink, smoke, or apply makeup while working in the laboratory.
7. If you cut or burn yourself while working, report this immediately to your laboratory instructor.
8. Broken glassware should be immediately brought to the attention of your laboratory instructor. Broken glass should be placed in a special sharps container for disposal and not in the trash container.
9. If using a Bunsen burner, tie back long hair. Do not lean over the countertop. When in use, always be aware of the flame. Keep flammable items away from the flame. Turn off the burner when not in use.
10. Before leaving the lab, make sure all items have been returned to their appropriate location.
11. After your work area is clear, wipe down your countertop with disinfectant before leaving.
12. Wash your hands thoroughly with antibacterial soap before leaving the lab.
13. Do not remove any item from the lab unless you have been directed to do so by the laboratory instructor.

Guidelines for Working with Biosafety Level (BSL) 1 Bacteria

Handling live bacteria in the laboratory, even those considered nonpathogenic, requires special guidelines beyond the general guidelines already mentioned. All bacteria are potentially pathogenic, especially if they gain entry into the human body. So observe the following guidelines when handling the biosafety level (BSL) 1 bacteria listed in the summary table.

1. Do not put anything into your mouth when working with cultures. Do not pipette by mouth; use a pipette aid instead. Keep your hands, pencil, pen, etc., away from your mouth, eyes, and nose.
2. When inoculating cultures, sterilize the loop or needle before placing it on the counter.
3. Always keep tubes in test tube racks when working with liquid media. Do not stand them up or lay them down on the countertop where they may spill.
4. If you accidentally spill a culture, cover the spill with a paper towel, flood it with disinfectant, and notify your laboratory instructor.
5. Place all used culture media, paper towels, gloves, etc., into the waste container designated by your laboratory instructor. A separate waste container for sharps (slides, pipettes, swabs, broken glass, etc.) will also be provided. All this waste will be autoclaved before disposal or reuse. Do not throw any of these items into the trash container.
6. If you have a burn or wound on one of your hands, cover it with a plastic strip and wear disposable gloves for added protection.

Guidelines for Working with Biosafety Level (BSL) 2 Bacteria

Handling pathogenic bacteria in the laboratory requires special guidelines beyond the general guidelines and those for BSL 1 bacteria. The following additional guidelines apply when working with the biosafety level (BSL) 2 bacteria listed in the summary table.

1. When handling pathogens, access to the laboratory must be restricted to only those working in the lab.
2. Disposable gloves and a lab coat must be worn. The gloves should be disposed of in a container designated by the instructor. The lab coat must be removed before leaving and kept in a designated area of the lab.
3. Avoid creating aerosols when working with pathogens. If there is a chance of creating tiny airborne droplets, work under a safety hood.

Summary of Biosafety Levels for Infectious Agents

Biosafety level (BSL)	Description of infectious agents	Examples from this lab manual
1	Agents that typically do not cause disease in healthy adults; they generally do not pose a disease risk to humans.	*Alcaligenes denitrificans* *Alcaligenes faecalis* *Bacillus cereus* *Bacillus subtilis* *Corynebacterium pseudodiphtheriticum* *Enterobacter aerogenes* *Enterococcus faecalis* *Escherichia coli* *Micrococcus luteus* *Neisseria sicca* *Proteus vulgaris* *Pseudomonas aeruginosa* *Serratia marcescens* *Staphylococcus epidermidis* *Staphylococcus saprophyticus*
2	Agents that can cause disease in healthy adults; they pose moderate disease risk to humans.	*Klebsiella pneumoniae* *Mycobacterium phlei* *Salmonella typhimurium* *Shigella flexneri* *Staphylococcus aureus* *Streptococcus pneumoniae* *Streptococcus pyogenes*
3	Agents that can cause disease in healthy adults; they are airborne and pose a more serious disease risk to humans.	None; these agents are not used in this lab manual.
4	Agents that can cause disease in healthy adults; they pose a lethal disease risk to humans; no vaccines or therapy available.	None; these agents are not used in this lab manual.

Universal Precautions

All human blood and certain other body fluids are treated as if they are infectious for blood-borne pathogens, such as human immunodeficiency virus (HIV), hepatitis B virus (HBV), and hepatitis C virus (HCV). Such precautions are the rule among nurses, doctors, phlebotomists, and clinical laboratory personnel, and are a critical component of infection control.

1. Wear gloves.
2. Change gloves when they are soiled or torn.
3. Remove gloves when you are finished handling a specimen, and before you touch other objects such as drawer handles, door knobs, refrigerator handles, pens/pencils, and paper.
4. Wash hands thoroughly with soap and water after removing gloves.
5. Dispose of gloves and blood-contaminated materials in a biohazard receptacle.

   Additional precautions that may not apply to this laboratory exercise:
6. Wear a lab coat when soiling with blood or body fluids is possible.
7. Wear a mask, goggles, or glasses with side shields if splashing of the face is possible.

Safety Commitment

I have read and understand the safety guidelines described above. I declare my commitment to safety in the microbiology laboratory and promise to follow each rule during the course of the semester.

_____   _____

Name   Date

Survey of Microscopic Organisms

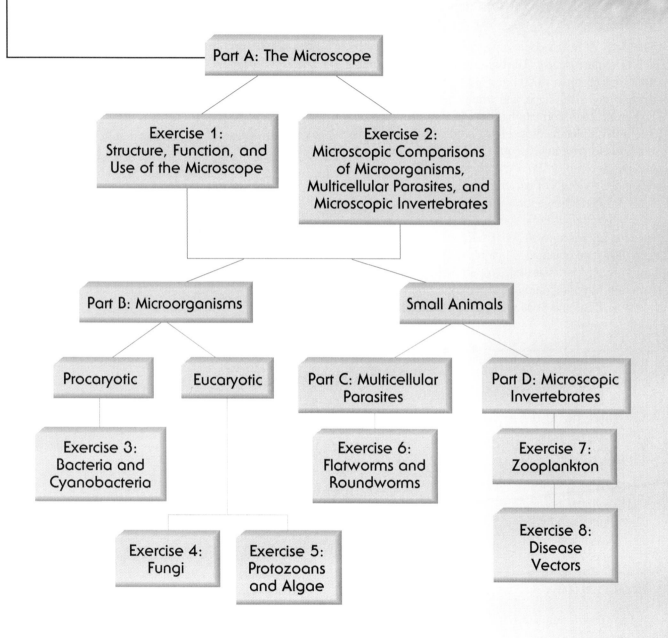

Part A: The Microscope

Exercise 1:
Structure, Function, and
Use of the Microscope

Exercise 2:
Microscopic Comparisons
of Microorganisms,
Multicellular Parasites, and
Microscopic Invertebrates

Part B: Microorganisms

Small Animals

Procaryotic

Eucaryotic

Part C: Multicellular
Parasites

Part D: Microscopic
Invertebrates

Exercise 3:
Bacteria and
Cyanobacteria

Exercise 6:
Flatworms and
Roundworms

Exercise 7:
Zooplankton

Exercise 4:
Fungi

Exercise 5:
Protozoans
and Algae

Exercise 8:
Disease
Vectors

1

Structure, Function, and Use of the Microscope

Background

The study of microscopic organisms is greatly aided by the use of microscopes. The **light microscope (LM)** magnifies objects up to 1,000 times (1,000×) and can be used to study cell size, shape, and arrangement. However, the LM gives little information about internal cell structures. The internal details of a cell are studied using a **transmission electron microscope (TEM)**, since useful magnifications of up to 100,000× are possible. The infection of a cell by viruses or bacteria can also be studied using a TEM. In addition, a three-dimensional view of cells in their natural environment is possible with a **scanning electron microscope (SEM)**. Useful magnifications of up to 20,000× are obtained with a SEM.

This exercise is designed to familiarize you with the structure, function, and use of the light microscope. In addition, TEM and SEM views of cells will be provided for comparison.

Materials

Prepared slides (2)
    Blood (human)
    Budding yeast

Equipment
    Microscope

Miscellaneous supplies
    Immersion oil
    Lens paper

Procedure

1. Familiarize yourself with the structure and function of the light microscope by reviewing the following: (a) the microscope in figure 1.1; (b) the parts of the microscope and their functions in table 1.1; and (c) the magnifications obtained using different objectives in table 1.2. Complete step 1 of the laboratory report.

Table 1.1 Functions of the Parts of the Light Microscope*

Part	Function
1. Ocular (eyepiece)	Magnifies image, usually 10x
2. Thumb wheel	Adjusts distance between oculars to match your eyes
3. Lock screw	Secures head after rotation
4. Head	Holds oculars
5. Arm	Holds head and stage
6. Revolving nosepiece	Rotates objective lenses into viewing position
7. Objective	Magnifies image, usually low (4×), medium (10×), high dry (40×), and oil-immersion (100×)
8. Slide holder	Fixed and movable parts secure slide on stage
9. Mechanical stage	Includes slide holder and is used to locate specimen
10. Stage	Holds slide
11. Stage aperture	Admits light
12. Condenser	Focuses light on specimen and fills lens with light
13. Diaphragm lever	Controls amount of light entering stage aperture
14. Substage-adjustment knob	Raises and lowers condenser
15. Mechanical-stage control	Moves slide back and forth on stage
16. Light source	Illuminates specimen
17. Coarse-adjustment knob	Rapidly brings specimen into focus
18. Fine-adjustment knob	Slowly brings specimen into sharp focus
19. Base	Supports microscope

*Parts are listed in order from top to bottom, and their numbers correspond to those in figure 1.1.

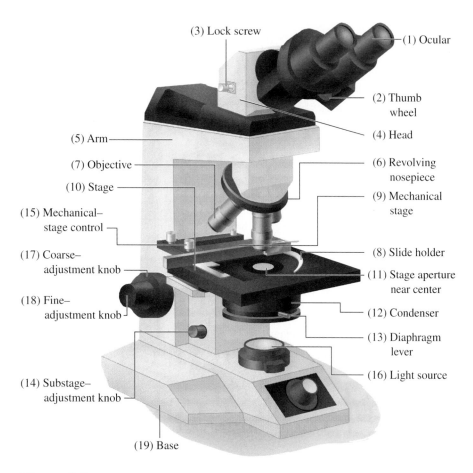

(3) Lock screw

(1) Ocular

(2) Thumb wheel

(4) Head

(5) Arm

(6) Revolving nosepiece

(7) Objective

(9) Mechanical stage

(10) Stage

(15) Mechanical–stage control

(8) Slide holder

(17) Coarse–adjustment knob

(11) Stage aperture near center

(18) Fine–adjustment knob

(12) Condenser

(13) Diaphragm lever

(16) Light source

(14) Substage–adjustment knob

(19) Base

Figure 1.1 The parts of the microscope.

Table 1.2 Total Magnification Possible with Different Objective Lenses of the Light Microscope			
Power	**Objective lens**	**Ocular lens**	**Total magnification**
Low	4×	10×	40×
Medium	10×	10×	100×
High dry	40×	10×	400×
Oil-immersion	100×	10×	1,000×

2. Table 1.3 lists the steps for using the light microscope. Follow these steps carefully as you examine two slides: human blood and budding yeast. Using figure 1.2 as a guide, identify as many of the cell types and structures as you can. For each slide, record in the laboratory report what you see at 40×, 100×, 400×, and 1,000×.

3. Examine the photographs of the TEM (figure 1.3) and the SEM (figure 1.4). Also examine the images of cells that these microscopes provide (figures 1.5–1.8). How do these views of cells differ from those provided by the light microscope?

Table 1.3 Steps in the Use of the Light Microscope

Carry the microscope upright with two hands (figure 1.9, p.10). Place the microscope on the countertop, plug it in, and turn on the light. Follow these steps as you examine the human blood and budding yeast slides:

1. Clip the slide into place on the stage using the slide holder.

2. Use the mechanical-stage control to move the slide so that the specimen is centered over the condenser.

3. Rotate the nosepiece to position the 4× objective (figure 1.10a, p. 11). When this objective is in place over the specimen, move the coarse-adjustment knob until the stage and objective are as close together as possible.

4. While looking through the oculars, move the coarse-adjustment knob to slowly increase the distance between the stage and the objective. Stop when the specimen comes into focus.

5. Adjust the distance of the ocular lens by moving the thumb wheel until two images become one.

6. Close your left eye, and focus for the right eye using the fine-adjustment knob. Close your right eye, and focus for the left eye using the focusing ring on the left ocular lens. Open both eyes and move the fine-adjustment knob until a sharp image is obtained. You are now ready to make your observations at 40× total magnification.

7. Center the specimen, and then rotate the nosepiece to position the 10× objective (figure 1.10b, p. 11). Since most microscopes are parfocal, the only adjustment that should be necessary is the fine adjustment. When the image is sharp, make your observations at 100× total magnification.

8. Rotate the nosepiece to position the 40× objective (figure 1.10c, p. 11). Move the fine-adjustment knob, and make your observations at 400× total magnification.

9. Move the 40× objective out of the way, and place a drop of immersion oil on top of the specimen. Position the 100× oil-immersion objective (figure 1.10d, p. 11). Move only the fine-adjustment knob. You may need to open the iris diaphragm with the diaphragm lever to allow more light to enter the objective lens. Make your observations at 1,000× total magnification.

10. When observations are complete, position the 4× objective lens and wipe the oil off the oil-immersion objective with a piece of lens paper. Remove the slide from the stage, and wipe off the oil if the specimen is covered by a coverslip. If not, let the oil drain off by placing the slide upright in a slide box.

11. When finished, turn off the light, unplug the cord, and wrap it around the base. Return the microscope to the storage cabinet.

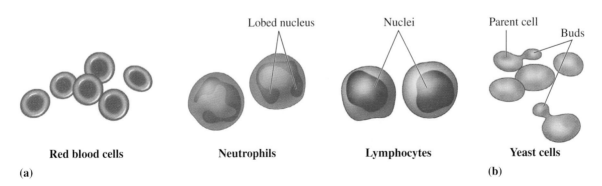

(a)

Lobed nucleus   Nuclei   Parent cell   Buds

Red blood cells   **Neutrophils**   **Lymphocytes**   **Yeast cells**

(b)

Figure 1.2 (a) Formed elements of human blood (1,000×); (b) Yeast cells (1,000×).

Figure 1.3 Transmission electron microscope (TEM).

Figure 1.4 Scanning electron microscope (SEM).

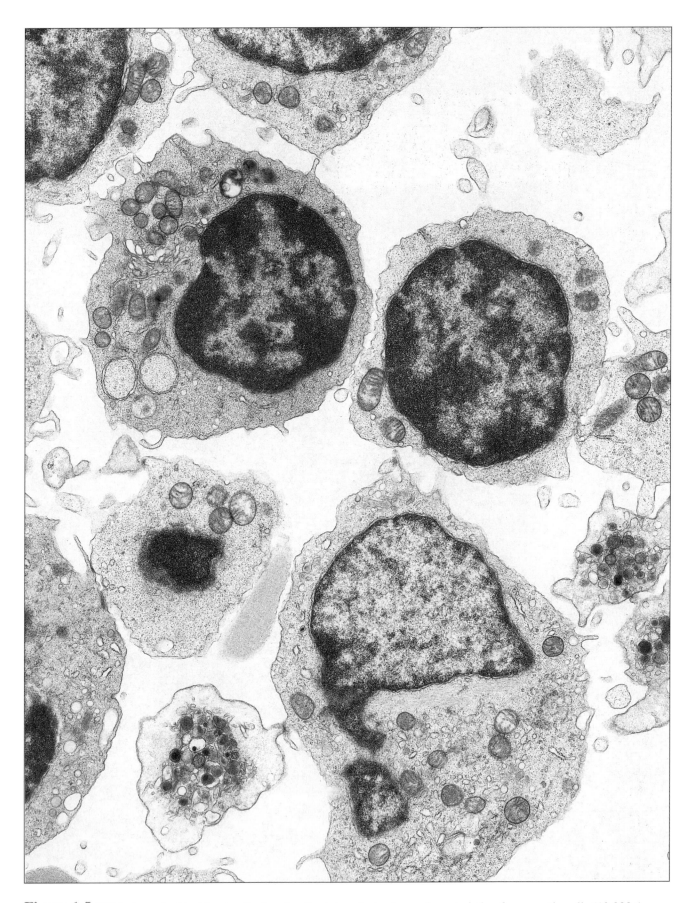

Figure 1.5 TEM view of white blood cells showing the internal structures characteristic of eucaryotic cells (12,000×).

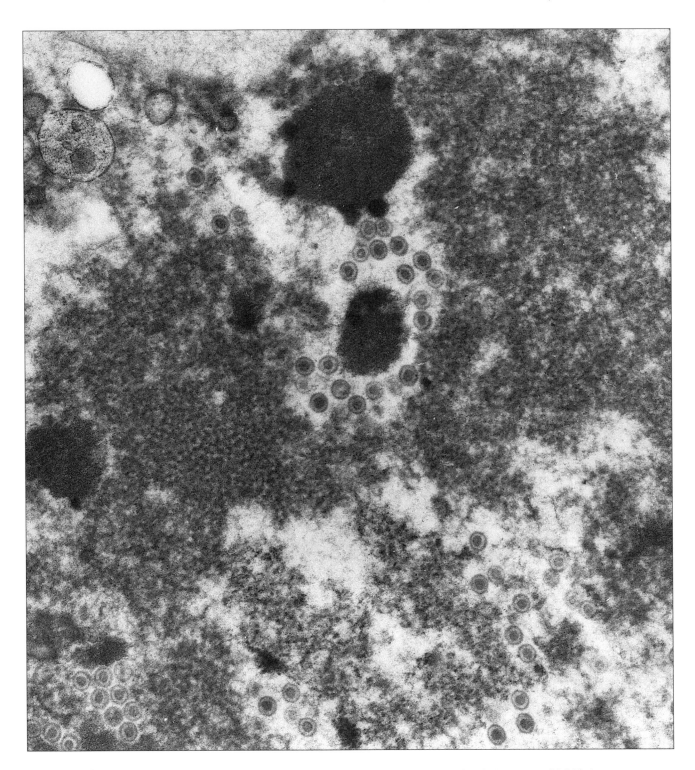

Figure 1.6 TEM view of a virus-infected cell. Viruses are the circular particles with dark centers (20,000×).

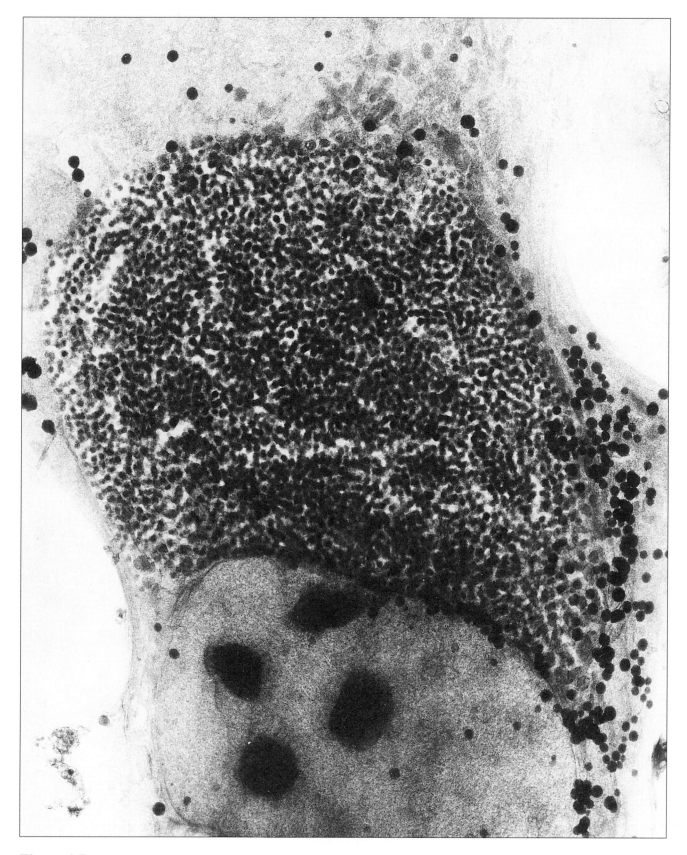

Figure 1.7 TEM view of a *Chlamydia*-infected cell. *Chlamydia* bacteria are the numerous dark circles (3,000×).

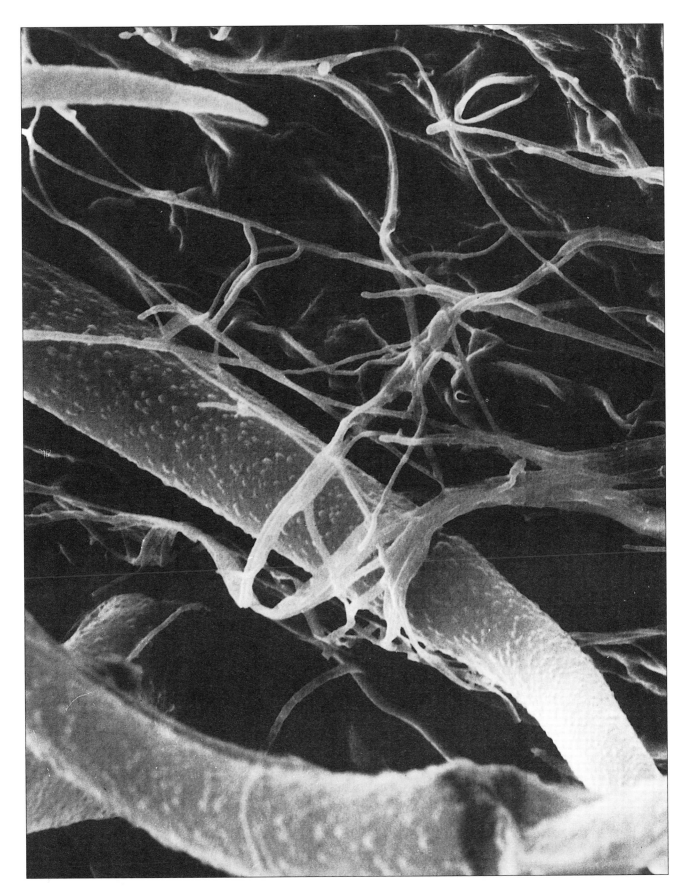

Figure 1.8 SEM view of fungal hyphae on the surface of a potato leaf (5,000×).

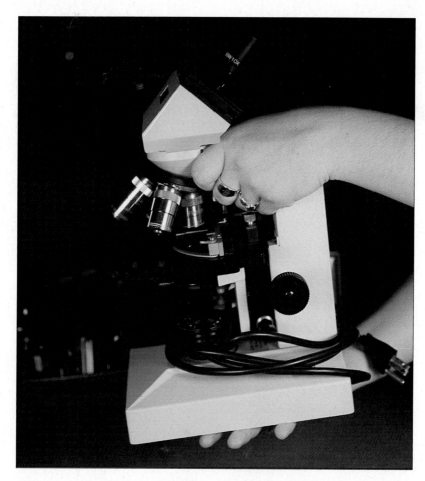

Figure 1.9 Method used to carry the light microscope.

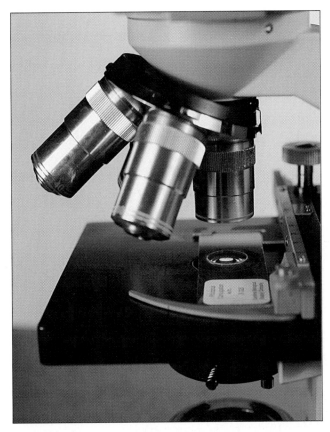

(a) 4× objective

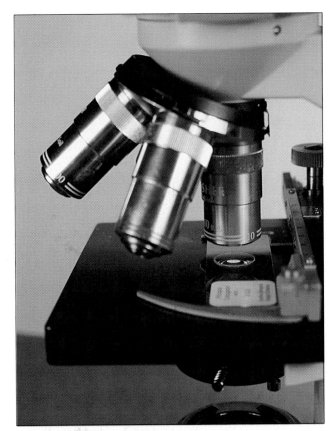

(b) 10× objective

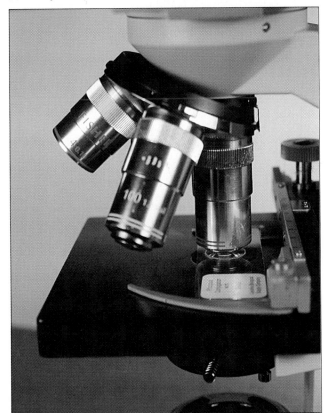

(c) 40× objective

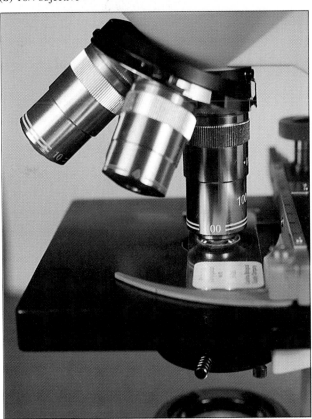

(d) 100× oil-immersion objective

Figure 1.10 Positions of light microscope objectives when viewing the specimen.

LABORATORY REPORT

NAME ———————————————————— DATE ——————————————

LAB SECTION ——————————————————————————————————

Structure, Function, and Use of the Microscope

1. Identify the parts (a–f) of the microscope below, and fill in their functions.

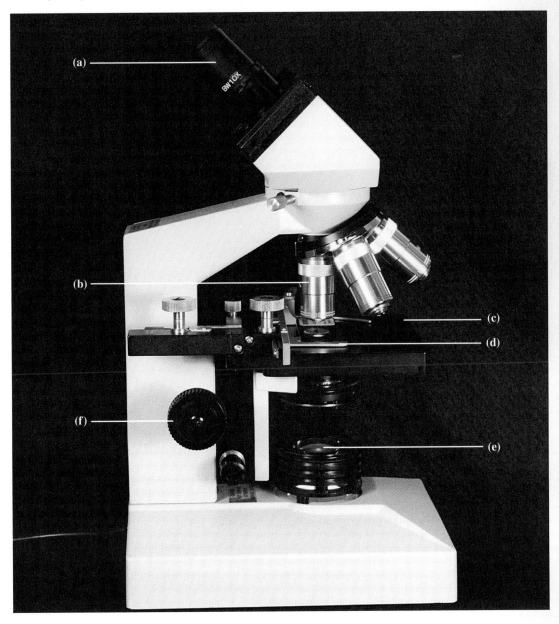

Part	Function	Part	Function
a. ——— ————————		d. ——— ————————	
b. ——— ————————		e. ——— ————————	
c. ——— ————————		f. ——— ————————	

2. Depict the morphology of a few representative cells at each total magnification. Try to draw the cells at the size scale you observed.

    a. **Human blood**
       Draw and label the
       cell types you find.

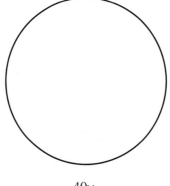

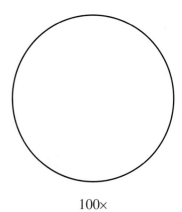

40×   100×

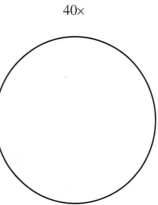

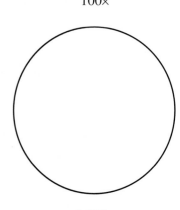

400×   1,000×

    b. **Budding yeast**
       Draw and label parent
       cells and buds you find.

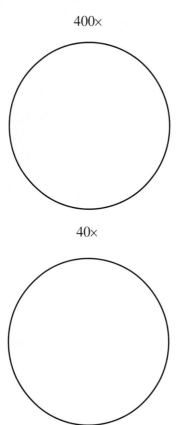

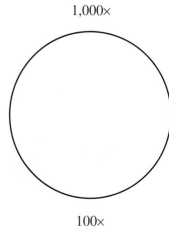

40×   100×

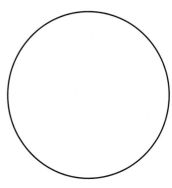

400×   1,000×

3. Which microscope (LM, TEM, or SEM) would be most useful to study the following?

    a. Size of cells _____

    b. Whether or not a cell has a nucleus (i.e., is procaryotic or eucaryotic) _____

    c. Whether or not a cell is infected with viruses _____

    d. A three-dimensional view of cells attached to a surface _____

    e. Cell shapes and arrangements _____

    f. Cells infected with *Chlamydia* _____

4. Answer the following questions in the space provided.

    a. (1) Give the general formula used to calculate the total magnification:

        _____ × _____ = total magnification

      (2) What is the total magnification when using the 100× oil-immersion objective lens? _____

    b. In general, should the condenser be kept close to or far from the stage? Explain.

    c. When increasing magnification from high dry to oil-immersion, should the iris diaphragm be open or closed? How is this done? Does this adjustment increase or decrease the light reaching the objective lens?

    d. Explain why oil must be used with the oil-immersion lens.

    e. Based on your observations of blood cells and yeast cells, which total magnification would you recommend for best viewing? Explain.

Microscopic Comparisons of Microorganisms, Multicellular Parasites, and Microscopic Invertebrates

Background

Microorganisms (bacteria, cyanobacteria, fungi, protozoans, and algae) and small animals (multicellular parasites and microscopic invertebrates) display a variety of shapes and sizes (table 2.1). Figure 2.1 depicts

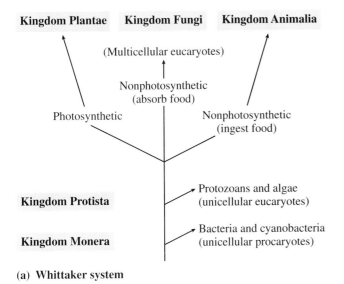

(a) Whittaker system

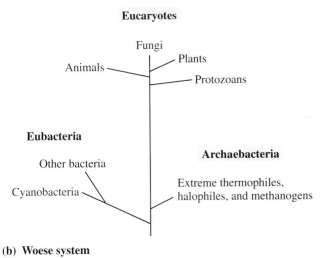

(b) Woese system

Figure 2.1 Two classification systems recognized by biologists and microbiologists: (a) the five-kingdom classification system of R. H. Whittaker; (b) the three-domain system of C. Woese.

two widely accepted classification systems for these organisms. The Whittaker system, which consists of five kingdoms, emphasizes differences in cellular traits and nutrition, while the Woese system, which consists of three domains, emphasizes differences in biochemical traits. Neither system includes the viruses, due to their unique makeup and method of replication.

In this exercise, you will use the microscope to make comparisons of the microscopic organisms examined in Section I. You will learn to make size measurements, and will measure a variety of microscopic organisms. After you measure, be sure to note the morphology of the microorganisms, multicellular parasites, and microscopic invertebrates.

Materials

Prepared slides (8)
  Select one slide from each category in
    table 2.1.

Equipment
  Light microscope

Miscellaneous supplies
  Immersion oil
  Lens paper
  Ocular micrometer
  Stage micrometer slide

Procedure

1. Clip the stage micrometer slide into position on the stage, and position the scale over the condenser (figure 2.2a, b). Focus on the scale using the 4× objective lens.

2. Align the ocular micrometer and stage micrometer scales as depicted in figure 2.2c. Now follow figure 2.2d to calibrate the ocular micrometer for the 4× objective lens.

17

Table 2.1 Typical Sizes of Selected Microscopic Organisms

Microscopic Organism	Size (in microns, μ)
Bacteria	
Bacillus	8
Escherichia coli	2-3
Spirillum	20
Staphylococcus	1
Treponema pallidum	15
Cyanobacteria	
Oscillatoria (filament)	400
Yeasts (fungi)	
Saccharomyces (with bud)	10
Molds (fungi)	
Aspergillus (conidiophore)	1,200
Rhizopus (zygospore)	400
Protozoans	
Amoeba proteus	300
Paramecium caudatum	200
Algae	
Diatoms (centric)	100
Diatoms (pennate)	50
Dinoflagellates	100
Spirogyra (filament)	2,500
Volvox (colony)	200
Multicellular parasites	
Clonorchis sinensis (liver fluke)	7,500
Dipylidium caninum (tapeworm proglottid)	2,500
Microscopic invertebrates	
Cyclops	500
Daphnia	500
Nauplius larvae	600
Tick	2,500

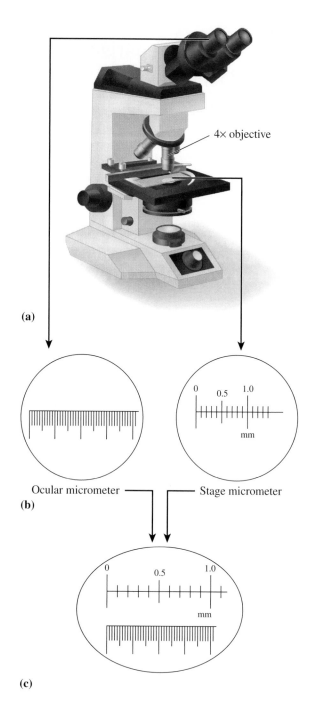

(a)

(b)

Ocular micrometer — Stage micrometer

(c)

Sample calculation from (c):

Stage micrometer 40×:	Ocular micrometer	Calibration
(1) 0.5 mm	20 ocular units (ou's)	0.025 mm/ou
(2) 1.0 mm	40 ocular units (ou's)	0.025 mm/ou
		Average = 25 μ/ou

(d)

Figure 2.2 Calibration of the ocular micrometer.

Table 2.2 Calculations in the Calibration of the Ocular Micrometer

Stage micrometer	Ocular micrometer		Calibration
a. 40×			
1. _____	_____		_____
2. _____	_____		_____
		Average	_____
b. 100×			
1. _____	_____		_____
2. _____	_____		_____
		Average	_____
c. 400×			
1. _____	_____		_____
2. _____	_____		_____
		Average	_____
d. 1,000×			
Calibration at 100× /10			_____

3. Repeat the calibration steps for the 10× and 40× objectives. To calculate the calibration for the 100× objective, take the calibration for the 10× objective and divide by 10. Record your ocular calibration results in table 2.2 and in the laboratory report.

4. Select one slide from each category in table 2.1 (eight total). Using your ocular calibration results, calculate and record in the laboratory report the size of each organism at the appropriate magnification. When comparing your results to those in table 2.1, do not expect results for every organism to be exactly like those shown, since the size of individual cells and cell groupings may vary.

5. Also be sure to depict the morphology of each organism in the circles provided in the laboratory report.

LABORATORY REPORT

NAME _____ DATE _____

LAB SECTION _____

Microscopic Comparisons of Microorganisms, Multicellular Parasites, and Microscopic Invertebrates

1. Record your ocular calibration results from table 2.2.

    40×: _____ µ/ocular unit (ou)

    100×: _____ µ/ou

    400×: _____ µ/ou

    1,000×: _____ µ/ou

2. Determine the size of each of the eight selected organisms by multiplying the length you measured in ocular units by the appropriate ocular calibration result recorded in question 1. Also sketch each organism in the circle provided.

Bacteria

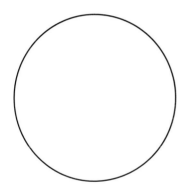

Organism _____

Magnification _____

Length (in ou's) _____

Size (_____ ou's × _____ µ/ou) = _____ µ

Cyanobacteria

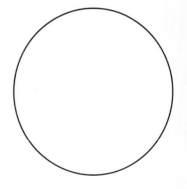

Organism _____

Magnification _____

Length (in ou's) _____

Size (_____ ou's × _____ µ/ou) = _____ µ

Yeasts (fungi)

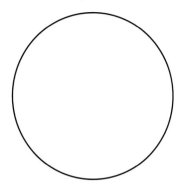

Organism _____

Magnification _____

Length (in ou's) _____

Size (_____ ou's × _____ μ/ou) = _____ μ

Molds (fungi)

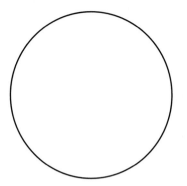

Organism _____

Magnification _____

Length (in ou's) _____

Size (_____ ou's × _____ μ/ou) = _____ μ

Protozoans

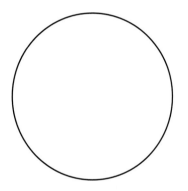

Organism _____

Magnification _____

Length (in ou's) _____

Size (_____ ou's × _____ μ/ou) = _____ μ

Algae

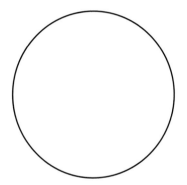

Organism _____

Magnification _____

Length (in ou's) _____

Size (_____ ou's × _____ μ/ou) = _____ μ

Multicellular parasites

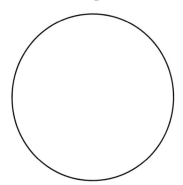

Organism ————————————

Magnification ————————

Length (in ou's) ————————

Size (————— ou's × ————— μ/ou) = ————— μ

Microscopic invertebrates

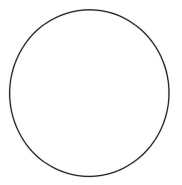

Organism ————————————

Magnification ————————

Length (in ou's) ————————

Size (————— ou's × ————— μ/ou) = ————— μ

3. List the eight organisms based on size, from smallest (1) to largest (8). Also list the magnification used to view each organism.

Organism	Size (μ)	Magnification used for viewing
1. ————————————	————————	————————
2. ————————————	————————	————————
3. ————————————	————————	————————
4. ————————————	————————	————————
5. ————————————	————————	————————
6. ————————————	————————	————————
7. ————————————	————————	————————
8. ————————————	————————	————————

4. Answer the following questions in the space provided.

   a. Based on your measurements and morphological observations, describe how the following microorganisms are different from one another.

   How are cyanobacteria different from bacteria? ————————————————

   How are yeasts different from bacteria? ————————————————

   How are molds different from bacteria and yeasts? ————————————————

   How are protozoans different from bacteria? ————————————————

   How are protozoans different from algae? ————————————————

b. Based on the Whittaker system, to which kingdom do the following organisms belong?

Organism	Kingdom
Protozoans	_____
Yeasts	_____
Bacteria	_____
Microscopic invertebrates	_____
Algae	_____
Molds	_____
Multicellular parasites	_____
Cyanobacteria	_____

5. Identify each of the following photos as bacteria, cyanobacteria, yeasts, molds, protozoans, algae, a multicellular parasite, or a microscopic invertebrate.

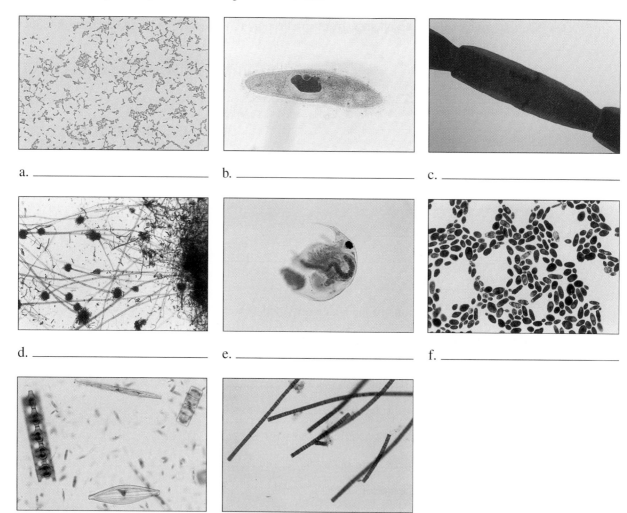

a. _____   b. _____   c. _____

d. _____   e. _____   f. _____

g. _____   h. _____

Microbial Procaryotes: Bacteria and Cyanobacteria

Background

Bacteria and cyanobacteria are both procaryotic microorganisms that belong to the Kingdom Monera in the Whittaker classification scheme. All pathogenic bacteria and most environmental bacteria are heterotrophic, lacking the light-absorbing pigments necessary to carry out photosynthesis. In contrast, cyanobacteria are autotrophic, containing the necessary light-absorbing pigments to carry out photosynthesis. Cyanobacteria and algae are responsible for the majority of the organic production that occurs in aquatic environments and wet soils.

Bacteria come in a variety of cell shapes, including rod, club, spirillum, spirochete, vibrio, and coccus (figure 3.1a). When bacteria grow (one cell dividing

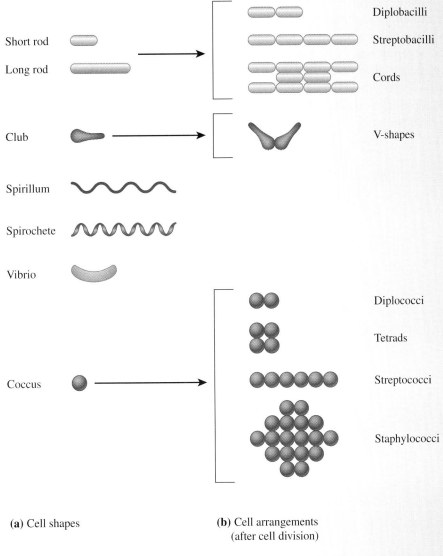

(a) Cell shapes

(b) Cell arrangements (after cell division)

Figure 3.1 (a) Cell shapes and (b) cell arrangements in bacteria.

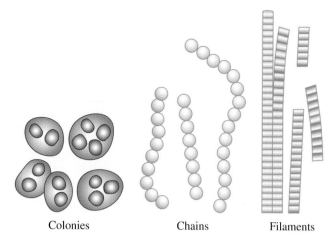

Figure 3.2 Cell arrangements in cyanobacteria.

to become two, the two cells dividing to become four, the four cells dividing to become eight, and so on), cells may separate or remain together. If cells remain together, a number of cell arrangements are possible, such as **diplobacilli, streptobacilli, cords, V-shapes, diplococci, tetrads, streptococci,** and **staphylococci** (figure 3.1b). Cell shape and arrangement are important characteristics used to identify bacteria.

Cyanobacteria come in a variety of shapes and arrangements as well. Their cells may be spherical or cubical, and arranged in a colony, chain, or filament (figure 3.2).

In this exercise, you will examine the variety of cell shapes and arrangements seen in bacteria and cyanobacteria.

Materials

Prepared slides
 Bacteria (11)
  Bacillus (large rods and streptobacilli)
  Corynebacterium diphtheriae (club and
   V-shapes); causes diphtheria
  Escherichia coli (short rods)
  Micrococcus luteus (cocci and tetrads)
  Mycobacterium tuberculosis (rods and
   cords); causes tuberculosis
  Neisseria gonorrhoeae (cocci and
   diplococci); causes gonorrhea
  Spirillum volutans (spirillum)
  Staphylococcus epidermidis (cocci and
   staphylococci)
  Streptococcus pyogenes (cocci and
   streptococci); causes strep throat
  Treponema pallidum (spirochete); causes
   syphilis
  Vibrio cholerae (vibrio); causes cholera

 Cyanobacteria (4)
  Anabaena (chains)
  Gleocapsa (colony)
  Nostoc (chains)
  Oscillatoria (filaments)

Equipment
 Light microscope

Miscellaneous supplies
 Immersion oil
 Lens paper

Procedure

1. Examine each of the prepared slides of bacteria
   using the oil-immersion lens. Note the variety
   of cell shapes and arrangements displayed
   by bacteria.

2. Examine the prepared slides of cyanobacteria
   using the 10× or 40× objective lens. Note the
   variety of forms displayed by cyanobacteria.

LABORATORY REPORT

NAME ——————————————— DATE ———————————————

LAB SECTION ———————————————————————————

Microbial Procaryotes: Bacteria and Cyanobacteria

1. Draw the bacteria and cyanobacteria you observed. Depict cell size, shape, and arrangement as accurately as possible.

   a. **Bacteria**

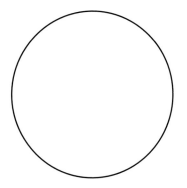

Bacillus

Magnification ——————

Cell shape ——————————————

Cell arrangement ——————————

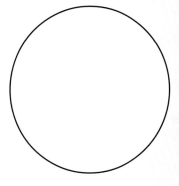

Corynebacterium diphtheriae

Magnification ——————

Cell shape ——————————————————

Cell arrangement ——————————————

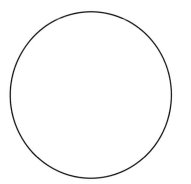

Escherichia coli

Magnification ——————

Cell shape ——————————————

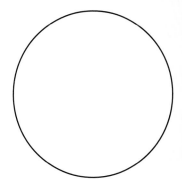

Micrococcus luteus

Magnification ——————

Cell shape ——————————————————

Cell arrangement ——————————————

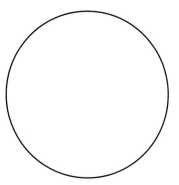

Mycobacterium tuberculosis

Magnification _____

Cell shape _____

Cell arrangement _____

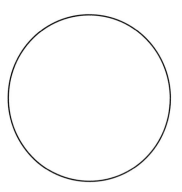

Neisseria gonorrhoeae

Magnification _____

Cell shape _____

Cell arrangement _____

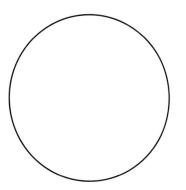

Spirillum volutans

Magnification _____

Cell shape _____

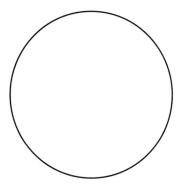

Staphylococcus epidermidis

Magnification _____

Cell shape _____

Cell arrangement _____

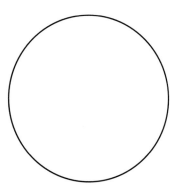

Streptococcus pyogenes

Magnification _____

Cell shape _____

Cell arrangement _____

Treponema pallidum

Magnification _____

Cell shape _____

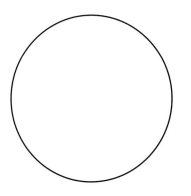

Vibrio cholerae

Magnification _____

Cell shape _____

b. **Cyanobacteria**

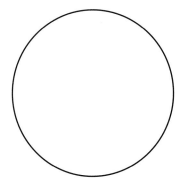

Anabaena

Magnification _____

Cell shape _____

Cell arrangement _____

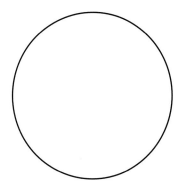

Gleocapsa

Magnification _____

Cell shape _____

Cell arrangement _____

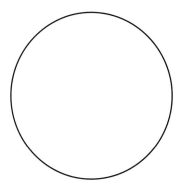

Nostoc

Magnification _____

Cell shape _____

Cell arrangement _____

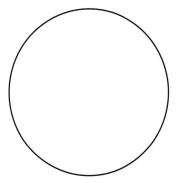

Oscillatoria

Magnification _____

Cell shape _____

Cell arrangement _____

2. Answer the following questions in the space provided.

    a. How are bacteria and cyanobacteria similar? Dissimilar?

    b. Why is "cyanobacteria" a more appropriate term than "blue-green algae"?

    c. Can cell shape and arrangement be useful in bacterial identification? If so, give three specific examples based on your observations.

3. Identify the cell shape and arrangement depicted in the following photographs of bacteria and cyanobacteria. Also give an example of a genus with these traits.

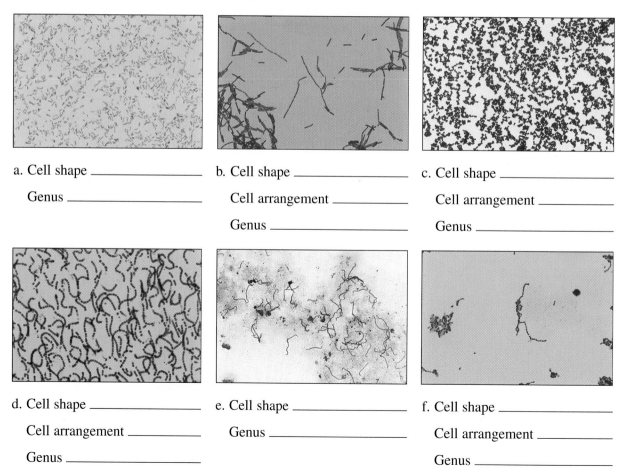

a. Cell shape _____

    Genus _____

b. Cell shape _____

    Cell arrangement _____

    Genus _____

c. Cell shape _____

    Cell arrangement _____

    Genus _____

d. Cell shape _____

    Cell arrangement _____

    Genus _____

e. Cell shape _____

    Genus _____

f. Cell shape _____

    Cell arrangement _____

    Genus _____

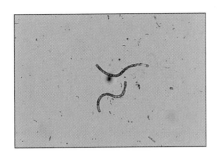 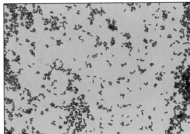 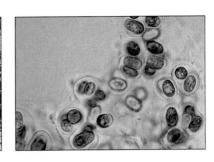

g. Cell shape _____

   Genus _____

h. Cell shape _____

   Cell arrangement _____

   Genus _____

i. Cell shape _____

   Cell arrangement _____

   Genus _____

Microbial Eucaryotes: Fungi

Background

Fungi exhibit a diversity of growth forms, such as **yeasts, molds, mushrooms, cup fungi,** and **lichens** (figure 4.1*a*). These organisms reproduce in a variety of ways: (1) formation of a **bud** from a parent yeast cell; (2) addition of new cells to chains of cells called **hyphae**; and (3) production of asexual and sexual spores (figure 4.1*b*). The type of sexual spore produced, whether **zygospore, ascospore,** or **basidiospore,** is used to classify fungi into groups.

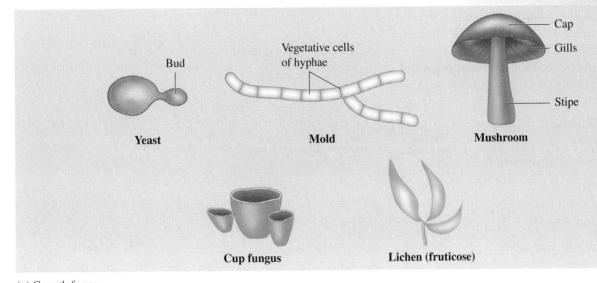

(a) Growth forms

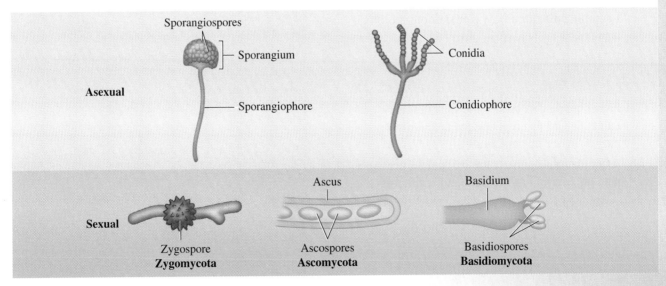

(b) Spore types

Figure 4.1 (a) Growth forms and (b) spore types in fungi.

This exercise will introduce you to the variety of growth forms in fungi and their methods of reproduction.

Materials

Fungal cultures on Sabouraud dextrose agar (4)
  Aspergillus (mold)
  Penicillium (mold)
  Rhizopus (mold)
  Saccharomyces (yeast)

Prepared slides of fungi (6)
  Candida albicans (pathogenic yeast); causes candidiasis
  Coprinus (mushroom with basidiospores on gills)
  Peziza (cup fungus with ascospores)
  Physcia (lichen with fungi and algae symbiosis)
  Rhizopus (bread mold with zygospores)
  Saccharomyces (brewing and baking yeast with buds)

Dry specimens of fungi obtained locally (2)
  Lichens (on a tree branch)
  Mushrooms (from a field or market)

Equipment
  Dissecting microscope
  Light microscope

Miscellaneous supplies
  Clear tape
  Glass slides
  Immersion oil
  Lactophenol cotton blue (for staining molds)
  Lens paper

Procedure

1. a. Examine the colonies of the four fungal cultures. The use of a dissecting microscope may aid your examination.

   b. After examining the colonies, make a pressure-tape preparation of the three mold cultures using the steps outlined in figure 4.2. Examine this preparation using the light microscope. Note the structures you see, including hyphae and asexual spores.

2. Examine the six prepared slides of fungi using the light microscope. Note the distinctive structure of each fungus examined, including hyphae, buds, conidia, zygospores, ascospores, and basidiospores.

3. Examine and record your observations of the dry specimens of a mushroom and lichens on a tree branch.

(a) Using a pipette, place a drop of lactophenol cotton blue on the center of the slide.

(b) Hold a piece of clear tape in a U-shape, sticky side down.

Sticky side

(c) Gently touch the surface of a mold colony.

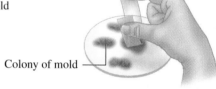

Colony of mold

(d) Place tape sticky side down in a drop of lactophenol cotton blue.

(e) Fold extra length of tape around edges of slide. Examine microscopically.

Figure 4.2 Pressure-tape preparation of fungi.

LABORATORY REPORT

NAME _____ DATE _____

LAB SECTION _____

Microbial Eucaryotes: Fungi

1. Record your results from the examination of fungal cultures.

Fungal culture	Colony description
Aspergillus (mold)	
Penicillium (mold)	
Rhizopus (mold)	
Saccharomyces (yeast)	

2. Draw from the microscopic examination of pressure-tape preparations of the three mold cultures.

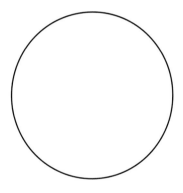

Aspergillus

Magnification _____

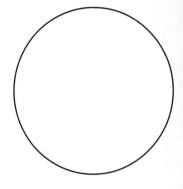

Penicillium

Magnification _____

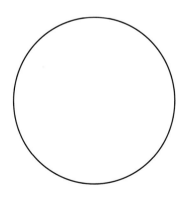

Rhizopus

Magnification _____

3. Draw the organisms you observed in the prepared slides of fungi.

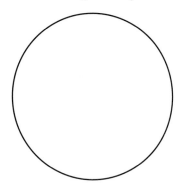

Candida albicans, pathogenic
yeast

Magnification _____

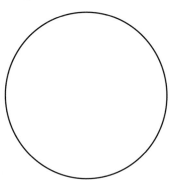

Coprinus, mushroom
(gill with basidiospores)

Magnification _____

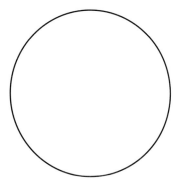

Peziza, cup fungus
(ascospores)

Magnification _____

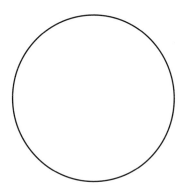

Physcia, lichen
(fungal filaments and algae)

Magnification _____

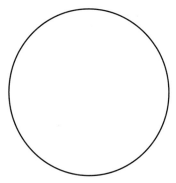

Rhizopus, bread mold
(zygospores)

Magnification _____

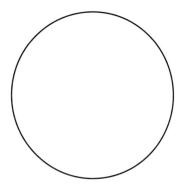

Saccharomyces, yeast
(cells with buds)

Magnification _____

4. a. Draw a mushroom, and label the following parts: stipe, cap, and gills.

    b. Draw a lichen on a tree branch. What two components form a lichen?

5. Answer the following questions in the space provided.

    a. Describe two differences between molds and yeasts.

    b. Name two characteristics that are used to distinguish one fungus from another.

    c. *Aspergillus fumigatus* causes an infection of the lungs called aspergillosis. How do you think this disease is acquired?

6. Identify the following photos.

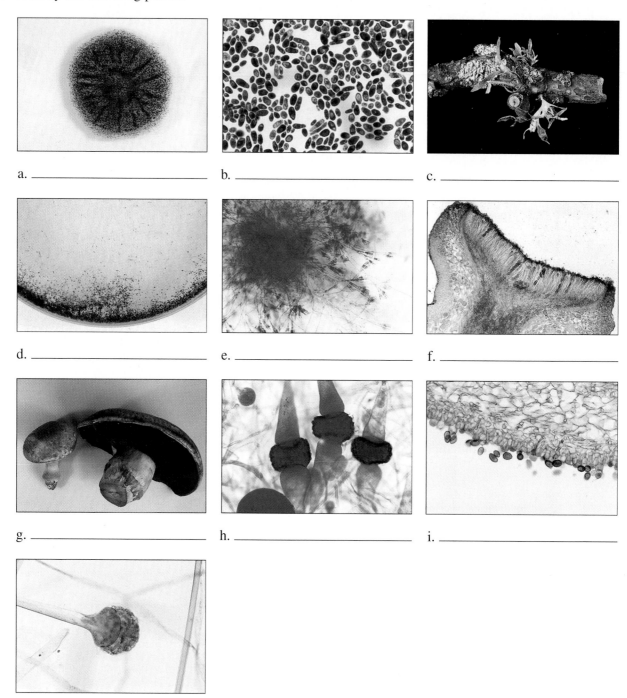

a. _____

b. _____

c. _____

d. _____

e. _____

f. _____

g. _____

h. _____

i. _____

j. _____

Microbial Eucaryotes: Protozoans and Algae

Background

In the Whittaker classification scheme, **protozoans** and **algae** are members of the **Kingdom Protista**. Protozoans are unicellular, nonphotosynthetic protists that are widespread in aquatic environments and wet soils. In this group, the type of organelle for motility is an important trait in classification. Protozoans have **pseudopodia, cilia,** or **flagella,** with the exception of the members of one group, the sporozoans, which do not have any of these structures (figure 5.1).

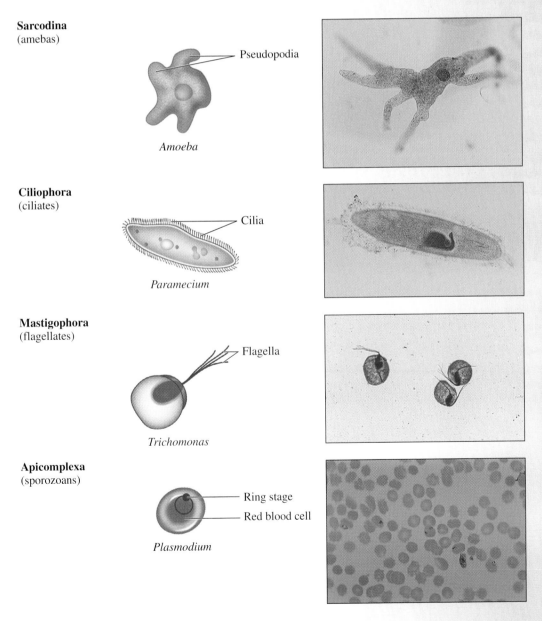

Sarcodina (amebas)

Pseudopodia

Amoeba

Ciliophora (ciliates)

Cilia

Paramecium

Mastigophora (flagellates)

Flagella

Trichomonas

Apicomplexa (sporozoans)

Ring stage

Red blood cell

Plasmodium

Figure 5.1 Representative protozoans, listed by phylum name.

Table 5.1 Distinguishing Traits of the Algae

Algal group (division)	Distinguishing traits
Diatoms (Chrysophyta)	Unicellular or chains of cells; silica cell walls consisting of overlapping halves; freshwater and marine
Dinoflagellates (Pyrrophyta)	Unicellular; armor of cellulose plates; 2 flagella; marine
Euglenoids (Euglenophyta)	Unicellular; red eyespot; 1 or 2 flagella; 2 to many chloroplasts; freshwater
Green algae (Chlorophyta)	Unicellular, colonial, and filamentous micro-algae; multicellular macro-algae; dominant pigment chlorophyll (green); freshwater and marine
Brown algae (Phaeophyta)	Multicellular macro-algae; dominant pigment fucoxanthin (brown); marine
Red algae (Rhodophyta)	Multicellular macro-algae; dominant pigment phycobilins (red); marine

Algae are photosynthetic protists that inhabit aquatic environments, where they are the primary agents responsible for the synthesis of organic molecules. They occur in a variety of forms, including unicellular, colonial, and filamentous micro-algae, and large, multicellular macro-algae. Several traits, such as morphology and photosynthetic pigments, are used to classify algae into the six groups shown in figure 5.2 and listed in table 5.1.

In this exercise, you will experience the diversity of the Kingdom Protista by examining a variety of protozoans and algae.

Materials

Prepared slides or live cultures
  Free-living protozoans (2)
    Amoeba (ameba)
    Paramecium (ciliate)

Prepared slides
  Pathogenic protozoans (2)
    Plasmodium (sporozoan); causes malaria
    Trichomonas vaginalis (flagellate); causes trichomoniasis

Micro-algae (6)
  Cladophora (filamentous green algae)
  Diatoms (unicellular and chain-forming chrysophytes)
  Dinoflagellates (unicellular pyrrophytes)
  Euglena (unicellular euglenoids)
  Spirogyra (filamentous green algae)
  Volvox (colonial green algae)

Preserved whole specimens of macro-algae (3)
  Padina (brown algae)
  Sargassum (brown algae)
  Ulva (green algae)

Pond water sample

Equipment
  Light microscope

Miscellaneous supplies
  Coverslips
  Glass slides
  Immersion oil
  Lens paper
  Pasteur pipette with bulb

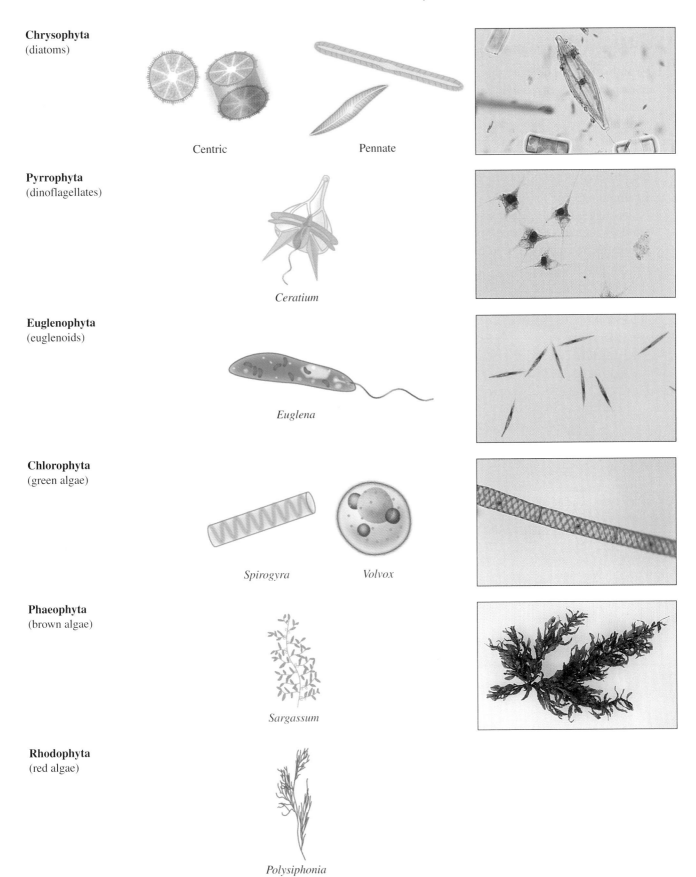

Figure 5.2 Representative algae, listed by division name.

Procedure

1. Examine microscopically the prepared slides or live specimens of free-living protozoans and the prepared slides of pathogenic protozoans. Live specimens of free-living protozoans can be examined using the wet mount preparation steps depicted in figure 5.3. For all protozoans, note the presence of pseudopodia, cilia, or flagella.

2. Examine microscopically the prepared slides of micro-algae. During your examination, note features such as cell morphology, cell arrangement, shape of chloroplasts, and unique structures.

3. Visually examine the preserved whole specimens of macro-algae, noting morphology, color, and unique structures.

4. Prepare several wet mounts of pond water using the procedure in figure 5.3. The pond water sample should contain both protozoans and micro-algae in the bottom sediment, so make sure you get some of this material with your pipette. Draw a few of the representative organisms you see under the microscope.

(a) Obtain water from the bottom of a live specimen container or pond water sample, and place on a glass slide.

(b) Using a pair of forceps, position a coverslip over the sample.

(c) Lower the coverslip over the sample. Remove excess water around the slip edge by blotting with tissue paper.

Figure 5.3 Wet mount preparation for viewing live specimens in pond water.

LABORATORY REPORT

NAME ————————————————— DATE —————————————

LAB SECTION ——————————————————————————

Microbial Eucaryotes: Protozoans and Algae

1. Draw your microscopic observations of free-living and pathogenic protozoans.

    a. **Free-living protozoans**

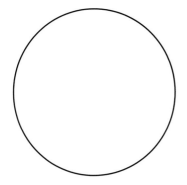

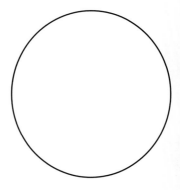

Amoeba (ameba)

Magnification ——————

Pseudopodia, cilia, or

flagella? ————————————————

Paramecium (ciliate)

Magnification ——————

Pseudopodia, cilia, or

flagella? ————————————————

    b. **Pathogenic protozoans**

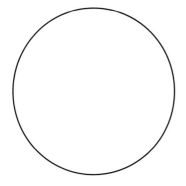

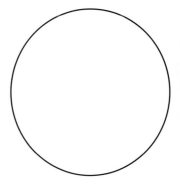

Plasmodium (sporozoan)

Magnification ——————

Pseudopodia, cilia, or

flagella? ——————————————

Trichomonas vaginalis (flagellate)

Magnification ——————

Pseudopodia, cilia, or

flagella? ——————————————

2. Sketch your microscopic observations of micro-algae.

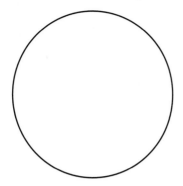

Cladophora (green algae)

Magnification _____

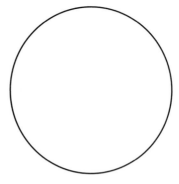

Diatoms (chrysophytes)

Magnification _____

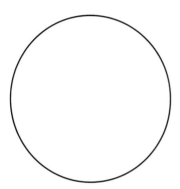

Dinoflagellates (pyrrophytes)

Magnification _____

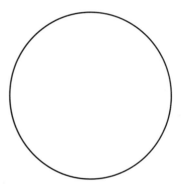

Euglena (euglenoid)

Magnification _____

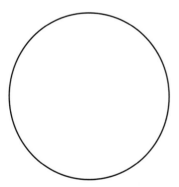

Spirogyra (green algae)

Magnification _____

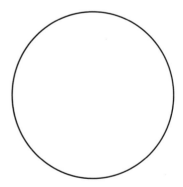

Volvox (green algae)

Magnification _____

3. Draw the general morphology of the preserved whole specimens of macro-algae.

Padina (brown algae)   *Sargassum* (brown algae)   *Ulva* (green algae)

4. Sketch several representative forms of protozoans and micro-algae you observed in the pond water.

   a. **Protozoans**

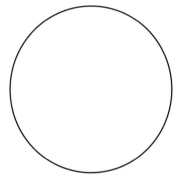

   Magnification _____

   b. **Micro-algae**

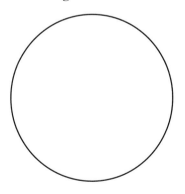

   Magnification _____

Magnification _____

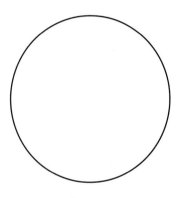

Magnification _____

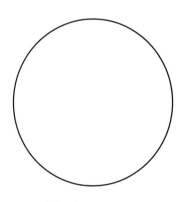

5. Answer the following questions in the space provided.

    a. On what basis are protozoan groups differentiated from one another?

    b. On what basis are algal groups differentiated from one another?

    c. How are protozoans and algae similar? Dissimilar?

6. Identify the following photos.

 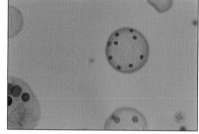 

a. Genus _____   b. Genus _____   c. Genus _____

    Algal group _____   Algal group _____   Algal group _____

Flatworms and Roundworms

Background

Multicellular parasites include flatworms (flukes and tapeworms) of the Phylum Platyhelminthes, and roundworms of the Phylum Nematoda. While flatworms are flattened in cross section, roundworms are round in cross section.

Flukes are flatworms with **oral suckers** and **ventral suckers**. Flukes, such as the Chinese liver fluke, *Clonorchis sinensis*, and the blood fluke, *Schistosoma mansoni*, infect humans, causing clonorchiasis and schistosomiasis, respectively. Clonorchiasis is acquired by ingesting raw or undercooked fish, while schistosomiasis is acquired when larvae penetrate human skin (figure 6.1).

Tapeworms are flatworms that have an anterior **scolex** for intestinal attachment and produce reproductive segments called **proglottids**. The tapeworms of the genus *Taenia* infect humans, causing taeniasis. This infection is contracted by ingesting undercooked beef or pork (figure 6.2).

Roundworms also infect humans, including the roundworm *Ascaris lumbricoides*, the cause of ascariasis; *Enterobius vermicularis*, the cause of enterobiasis; and *Trichinella spiralis*, the cause of trichinosis. Ascariasis and enterobiasis are contracted by ingesting food or water contaminated with roundworm eggs, while trichinosis is contracted by consuming undercooked pork (figure 6.3).

In this exercise, you will examine basic structural characteristics and life cycle aspects of these multicellular parasites.

Materials

Preserved specimen (1)
  Ascaris (roundworms, male and female)

Prepared slides
  Flukes (5)
    Clonorchis (adult)
    Clonorchis (eggs)
    Schistosoma (adult)
    Schistosoma (eggs)
    Schistosoma (cercaria)

  Tapeworms (4)
    Taenia (scolex)
    Taenia (mature or gravid proglottid)
    Taenia (eggs)
    Taenia (cysticercus)

  Roundworms (5)
    Ascaris (eggs)
    Enterobius (adult)
    Enterobius (eggs)
    Trichinella (adult)
    Trichinella (larvae)

Equipment
  Microscope

Miscellaneous supplies
  Immersion oil
  Lens paper

Procedure

1. Examine the prepared slides of flukes, noting unique structures, such as oral and ventral suckers.

2. Examine the prepared slides of tapeworms, noting unique structures, such as the scolex and reproductive proglottids.

3. a. Examine the prepared slides of the roundworms, noting unique structures.

   b. Examine the preserved specimens of *Ascaris lumbricoides*. Note the morphological differences between the male and female.

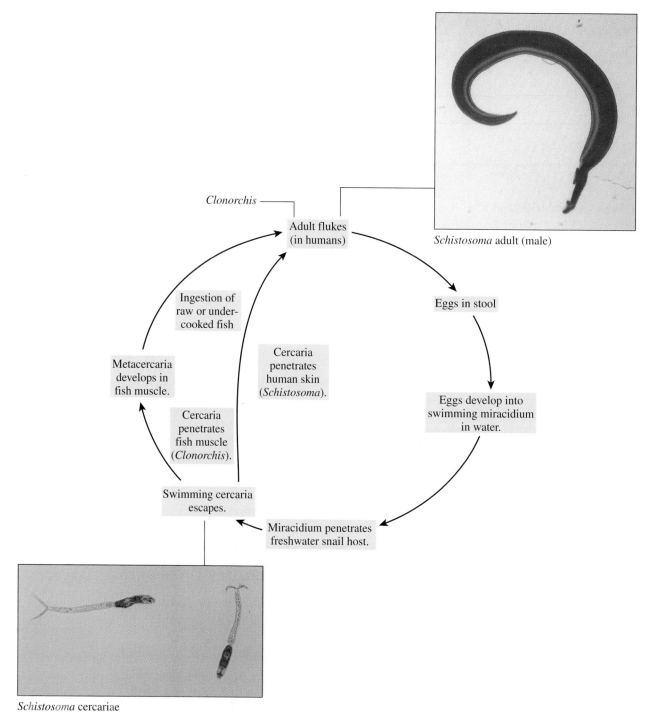

Schistosoma adult (male)

Clonorchis

Adult flukes
(in humans)

Ingestion of
raw or under-
cooked fish

Eggs in stool

Metacercaria
develops in
fish muscle.

Cercaria
penetrates
human skin
(*Schistosoma*).

Cercaria
penetrates
fish muscle
(*Clonorchis*).

Eggs develop into
swimming miracidium
in water.

Swimming cercaria
escapes.

Miracidium penetrates
freshwater snail host.

Schistosoma cercariae

Figure 6.1 Life cycles of two flukes, *Clonorchis sinensis* and *Schistosoma mansoni.*

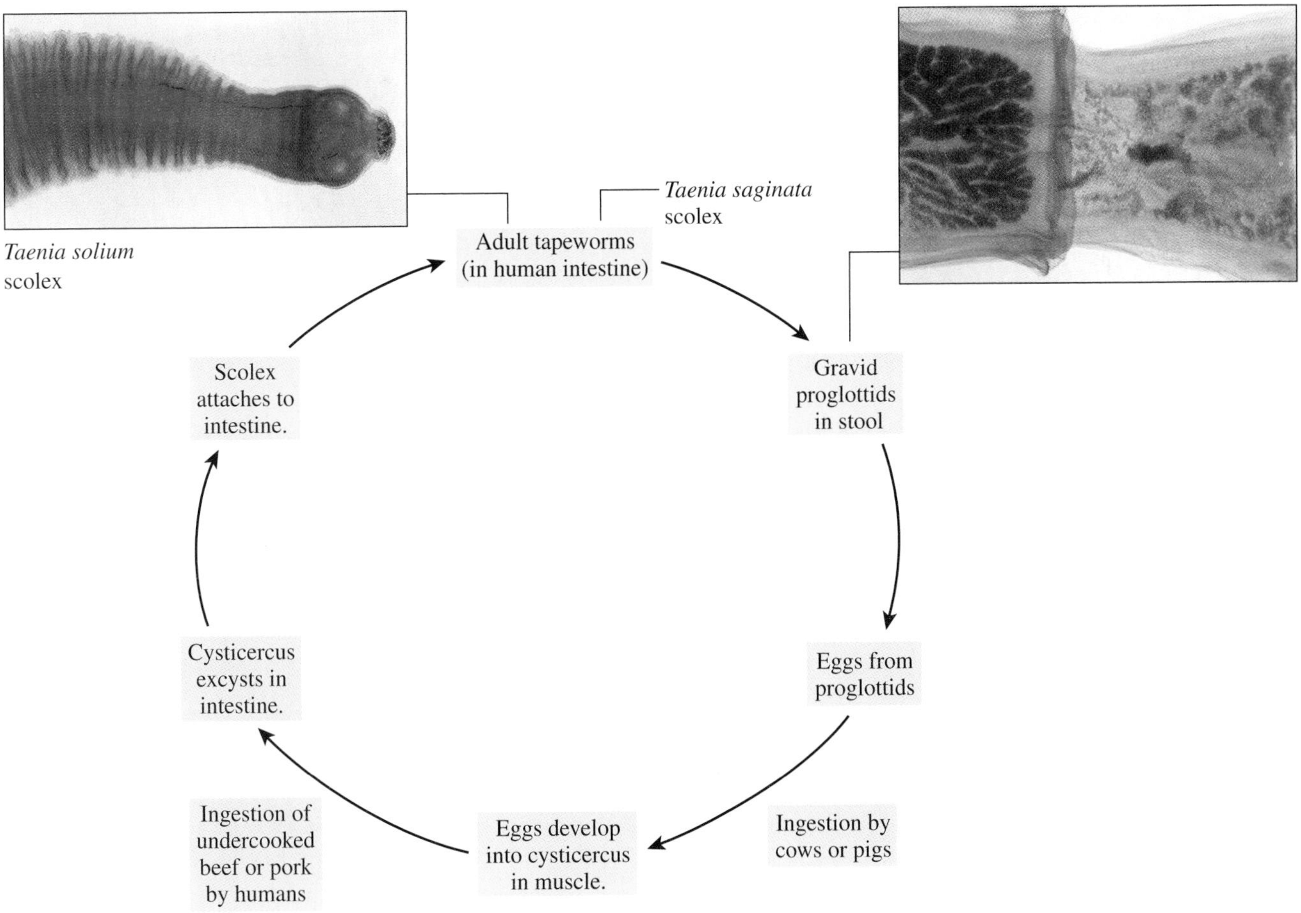

Taenia solium
scolex

Taenia saginata
scolex

Adult tapeworms
(in human intestine)

Scolex
attaches to
intestine.

Gravid
proglottids
in stool

Cysticercus
excysts in
intestine.

Eggs from
proglottids

Ingestion of
undercooked
beef or pork
by humans

Eggs develop
into cysticercus
in muscle.

Ingestion by
cows or pigs

Figure 6.2 Life cycle of the tapeworm, *Taenia.*

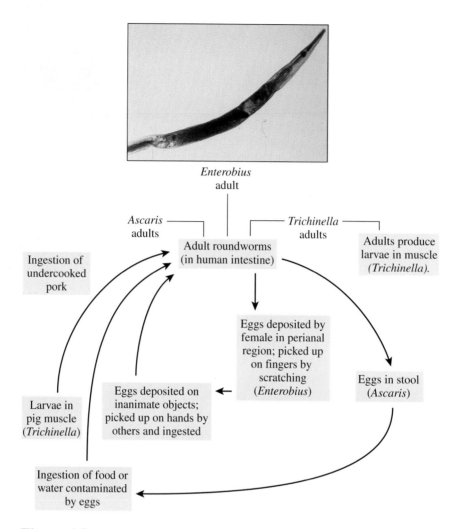

Figure 6.3 Life cycle of three roundworms.

LABORATORY REPORT

NAME _____ DATE _____

LAB SECTION _____

Flatworms and Roundworms

1. Sketch the specimens you examined.

   a. **Flatworms: Flukes**

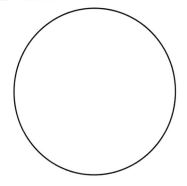

Clonorchis (adult)

Magnification _____

Clonorchis (eggs)

Magnification _____

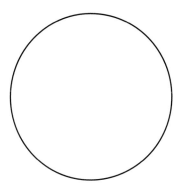

Schistosoma (adult)

Magnification _____

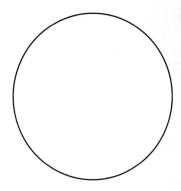

Schistosoma (eggs)

Magnification _____

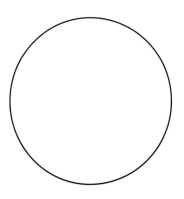

Schistosoma (cercaria)

Magnification _____

b. **Flatworms: Tapeworms**

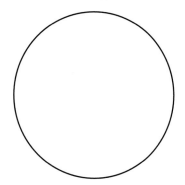

Taenia (scolex)

Magnification _____

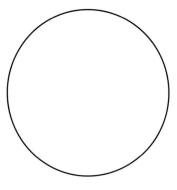

Taenia (proglottid, mature or gravid)

Magnification _____

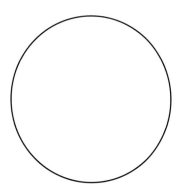

Taenia (eggs)

Magnification _____

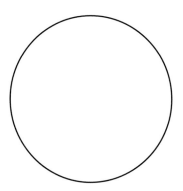

Taenia (cysticercus)

Magnification _____

c. **Roundworms**

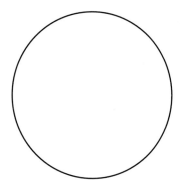

Ascaris (eggs)

Magnification _____

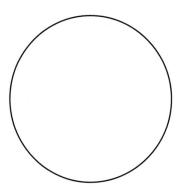

Enterobius (adult)

Magnification _____

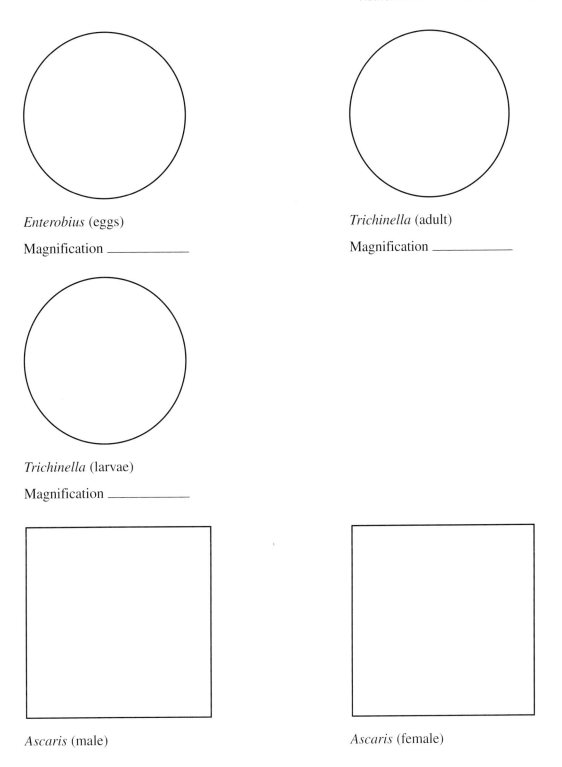

Enterobius (eggs)

Magnification _____

Trichinella (adult)

Magnification _____

Trichinella (larvae)

Magnification _____

Ascaris (male)

Ascaris (female)

2. Check the morphological features that apply.

Multicellular parasite	Body flattened in cross section	Body round in cross section	Oral/ventral suckers	Scolex with suckers/hooks	Reproductive segments (proglottids)
Flukes					
Tapeworms					
Roundworms					

3. Fill in this table.

Multicellular parasite	How Contracted?	Name of disease
Flukes Clonorchis		
Schistosoma		
Tapeworms Taenia		
Roundworms Ascaris		
Enterobius		
Trichinella		

4. Identify the following photos.

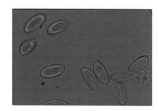  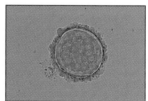 

a. _____   b. _____   c. _____   d. _____

Zooplankton

Background

A variety of **microscopic invertebrates** make up the small animal plankton, or zooplankton, of aquatic environments. **Zooplankton** feed on microscopic plant plankton, or **phytoplankton**, and in turn are fed upon by the higher-trophic-level consumers in aquatic food chains (figure 7.1). Zooplankton, therefore, are a critical food chain link in aquatic environments.

Zooplankton are subdivided into two categories, holoplankton and meroplankton. **Holoplankton** are the permanent members of the zooplankton, and include copepods, cladocerans, rotifers, and ostracods. **Meroplankton** are the temporary members of the zooplankton, and include the larval stages of benthic marine animals such as polychaetes, gastropods, barnacles, crabs, and starfish. Larvae change into adult animals and settle to the bottom to take on the benthic lifestyle.

In this exercise, you will examine prepared slides of representative zooplankton from freshwater and marine environments. Marine zooplankton will include examples of both holoplankton and meroplankton. You will also examine preserved samples of plankton, if these are available.

Materials

Prepared slides
  Freshwater zooplankton (3)
    Cyclops (copepod)
    Daphnia (cladoceran)
    Rotifers

Marine zooplankton (10)
    Bipinnaria (early starfish larvae)
    Brachiolaria (late starfish larvae)
    Calanus (copepod)
    Megalops (late crab larvae)
    Nauplius (early barnacle larvae)
    Ostracod
    Planula (early jellyfish larvae)
    Trochophore (early polychaete larvae)
    Veliger (gastropod larvae)
    Zoea (early crab larvae)

Preserved plankton samples (2)
    Freshwater
    Marine

Equipment
    Light microscope
    Dissecting microscope

Miscellaneous supplies
    Immersion oil
    Lens paper
    Pasteur pipette with bulb
    Sample dish

Procedure

1. Examine the prepared slides of freshwater zooplankton.

2. Examine the prepared slides of marine zooplankton, including both holoplankton and meroplankton.

3. Examine plankton samples, if available. Transfer some of the sample to a dish with a Pasteur pipette. Examine the contents in the dish with a dissecting microscope, and note the types of zooplankton you see.

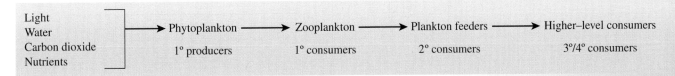

Figure 7.1 Food chain diagram of aquatic ecosystems.

LABORATORY REPORT

NAME ———————————————— DATE ————————————

LAB SECTION ——————————————————————————

Zooplankton

1. Draw the freshwater zooplankton you observed in the prepared slides.

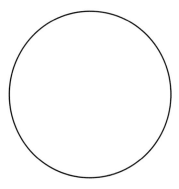

Cyclops (copepod)

Magnification ———————

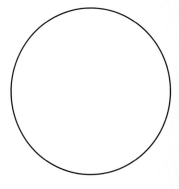

Daphnia (cladoceran)

Magnification ———————

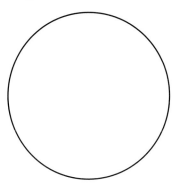

Rotifers

Magnification ———————

2. Draw the marine zooplankton you observed in the prepared slides.

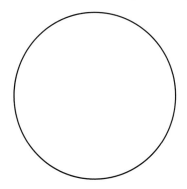

Bipinnaria (starfish larva)

Magnification ———————

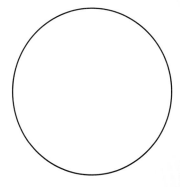

Brachiolaria (starfish larva)

Magnification ———————

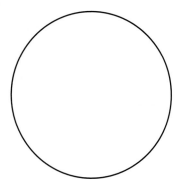

Calanus (copepod)

Magnification _____

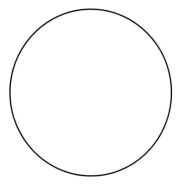

Megalops (crab larva)

Magnification _____

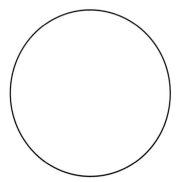

Nauplius (barnacle larva)

Magnification _____

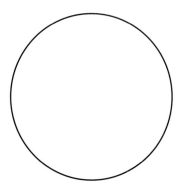

Ostracod

Magnification _____

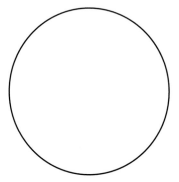

Planula (jellyfish larva)

Magnification _____

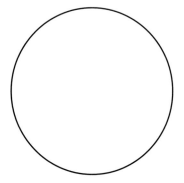

Trochophore (polychaete larva)

Magnification _____

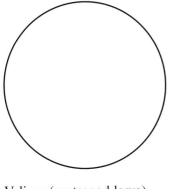

Veliger (gastropod larva)

Magnification _____

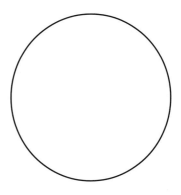

Zoea (crab larva)

Magnification _____

3. Draw several of the common zooplankton organisms you observed in the plankton samples.

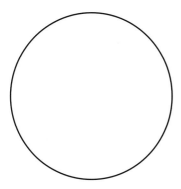

Freshwater plankton

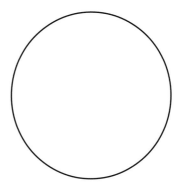

Marine plankton

4. Answer the following questions in the space provided.

   a. Depict an aquatic food chain showing the position of the zooplankton examined in this exercise.

   b. Explain the difference between holoplankton and meroplankton.

5. Identify the following members of the zooplankton.

a. _____

b. _____

c. _____

d. _____

e. _____

f. _____

g. _____

h. _____

i. _____

j. _____

Disease Vectors

Background

Disease-causing microorganisms can be transmitted in a variety of ways, including through air, food and water, and sexual contact. **Vectors** also transmit disease-causing microorganisms.

Most disease vectors are arachnids or insects that belong to the Phylum Arthropoda. Arthropod vectors include ticks, lice, mosquitoes, and fleas (figure 8.1). These organisms bite humans and in the process transmit pathogens.

In this exercise, you will examine the arthropod vectors of human diseases.

Materials

Prepared slides (10)
   Aedes (mosquito)
   Anopheles (mosquito)
   Chrysops (deer fly)
   Culex (mosquito)
   Dermacentor (tick)
   Glossina (tsetse fly)
   Ixodes (tick)
   Ornithodorus (tick)
   Pediculus (human louse)
   Xenopsylla (rat flea)

Equipment
   Light microscope
   Dissecting microscope

Procedure

Examine the prepared slides of arthropod vectors, noting their size, distinguishing structures, and unique features.

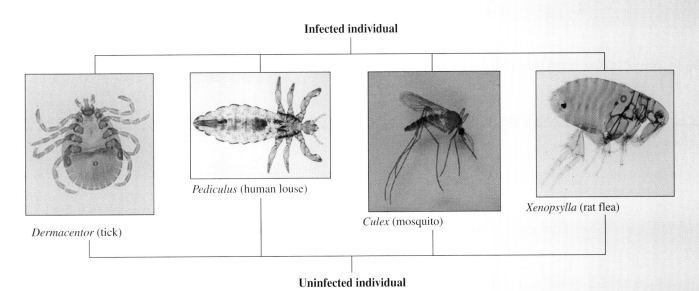

Infected individual

Pediculus (human louse)

Dermacentor (tick)

Culex (mosquito)

Xenopsylla (rat flea)

Uninfected individual

Figure 8.1 Selected examples of arthropod disease vectors.

LABORATORY REPORT

NAME _____ DATE _____

LAB SECTION _____

Disease Vectors

1. Draw the organisms you observed in the prepared slides.

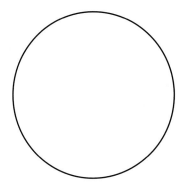

Aedes (mosquito)

Disease transmitted _____

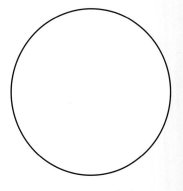

Anopheles (mosquito)

Disease transmitted _____

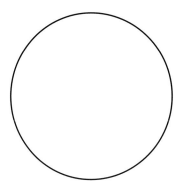

Chrysops (deer fly)

Disease transmitted _____

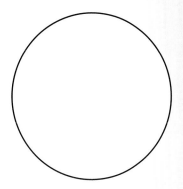

Culex (mosquito)

Disease transmitted _____

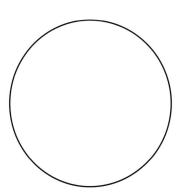

Dermacentor (tick)

Disease transmitted _____

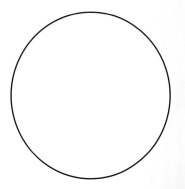

Glossina (tsetse fly)

Disease transmitted _____

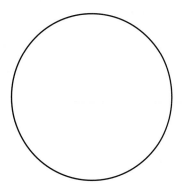

Ixodes (tick)

Disease transmitted _____

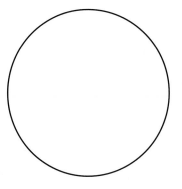

Ornithodorus (tick)

Disease transmitted _____

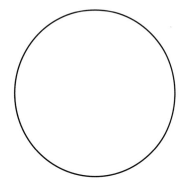

Pediculus (human louse)

Disease transmitted _____

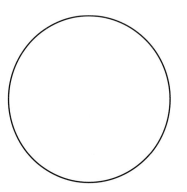

Xenopsylla (rat flea)

Disease transmitted _____

2. Answer the following questions in the space provided.

   a. Explain how these organisms transmit diseases.

   b. Explain why certain diseases transmitted by vectors, such as Lyme disease, occur more frequently in certain areas.

   c. *Ixodes*, the tick vector of Lyme disease, can be found attached to the skin after a walk in the woods. What would you recommend to a person going to the woods? What would you recommend to a person returning from the woods?

Staining Techniques

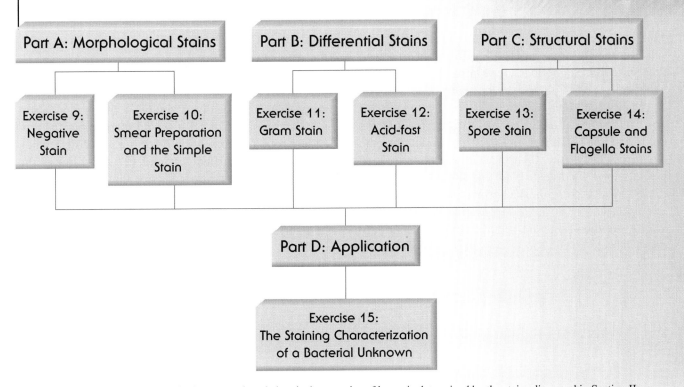

Part A: Morphological Stains

Exercise 9:
Negative
Stain

Exercise 10:
Smear Preparation
and the Simple
Stain

Part B: Differential Stains

Exercise 11:
Gram Stain

Exercise 12:
Acid-fast
Stain

Part C: Structural Stains

Exercise 13:
Spore Stain

Exercise 14:
Capsule and
Flagella Stains

Part D: Application

Exercise 15:
The Staining Characterization
of a Bacterial Unknown

Overview of the morphological, structural, and chemical properties of bacteria determined by the stains discussed in Section II.

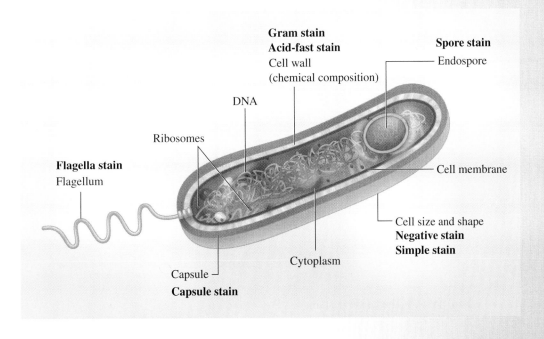

Gram stain
Acid-fast stain
Cell wall
(chemical composition)

DNA

Ribosomes

Flagella stain
Flagellum

Spore stain
Endospore

Cell membrane

Cell size and shape
Negative stain
Simple stain

Cytoplasm

Capsule
Capsule stain

9

Negative Stain

Background

Morphological stains color either bacterial cells themselves or their backgrounds to allow a clear microscopic view of cells. Such clear views provide information about cell size, shape, and arrangement.

Stains that color bacterial cells themselves carry a positive charge and are called **basic stains**. Basic stains color bacterial cells because they are attracted to the negatively charged cell surface. Basic stains include crystal violet, methylene blue, and safranin.

Stains that color the background surrounding bacterial cells carry a negative charge and are called **acidic stains**. Acidic stains are repelled by the negatively charged bacterial cell surface and, hence, color only the background (figure 9.1). However, this still provides a clear microscopic view, because the bacterial cells are seen in outline. Acidic stains include congo red, nigrosin, and india ink.

A single acidic stain used to color the background around cells is called a **negative stain**. There are two advantages of a negative stain: (1) it allows more accurate determination of cell size and shape, since the procedure requires no heating or staining of cells (which can cause cell shrinkage); and (2) it facilitates the microscopic observation of cells that are difficult to stain, such as spirilli and spirochetes.

In this exercise, you will use a single acidic stain to determine the cell morphology of several bacterial cultures.

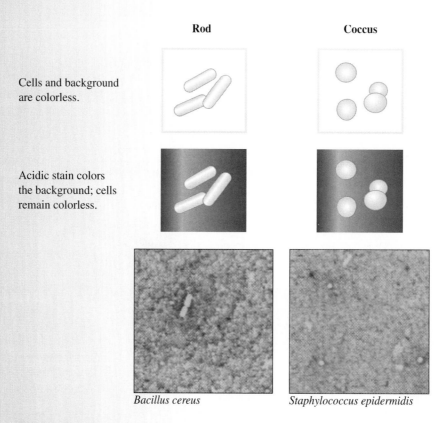

Rod   **Coccus**

Cells and background are colorless.

Acidic stain colors the background; cells remain colorless.

Bacillus cereus   *Staphylococcus epidermidis*

Figure 9.1 The negative stain.

Materials

Cultures (24–48-hour broth)
 Bacillus cereus (rod)
 Staphylococcus epidermidis (coccus)

Stains
 Nigrosin, india ink, or congo red

Equipment
 Light microscope

Miscellaneous supplies
 Bunsen burner and striker
 Disposable gloves (optional)
 Glass slides
 Immersion oil
 Inoculating loop
 Lens paper
 Wax pencil

Procedure

1. Place a drop of nigrosin, india ink, or congo red near the edge of a *clean* glass slide.

2. Aseptically obtain a loopful of a broth culture of *Bacillus cereus* by following the steps in figure 9.2.

3. Transfer the loopful of culture to the drop of stain on the slide, and mix the culture into the drop, as shown in figure 9.3*a, b*. **Always flame your loop before setting it down!**

4. Follow steps *c–e* in figure 9.3 to complete your preparation of a negative stain of *Bacillus cereus*.

5. Repeat steps 1–4 to prepare a negative stain of a broth culture of *Staphylococcus epidermidis*.

6. After you have completed both negative stains, examine them using the oil-immersion objective.

(a) Shake the culture tube from side to side to suspend the organisms.

(b) Heat the loop and wire to red-hot.

(c) Remove the cap, and flame the opening of the tube. Do not place the cap down on the table.

(d) After allowing the loop to cool for 5 seconds, remove a loopful of organisms. Avoid touching the sides of the tube.

(e) Flame the mouth of the tube again.

(f) Return the cap to the tube, and place the tube in a test tube rack. Transfer the loopful of organisms.

Figure 9.2 Aseptic procedure for removing an organism from a broth culture.

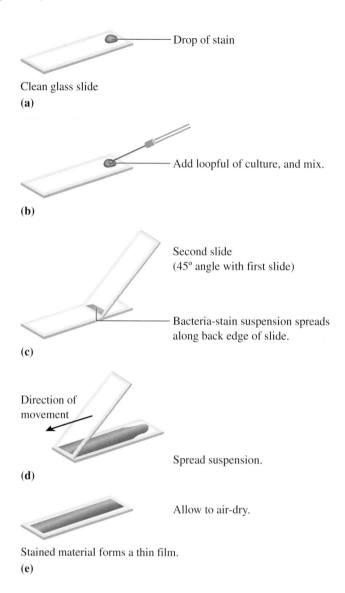

Drop of stain

Clean glass slide
(a)

Add loopful of culture, and mix.

(b)

Second slide
(45° angle with first slide)

Bacteria-stain suspension spreads
along back edge of slide.

(c)

Direction of
movement

Spread suspension.

(d)

Allow to air-dry.

Stained material forms a thin film.
(e)

Figure 9.3 Negative staining procedure.

LABORATORY REPORT

NAME _____ DATE _____

LAB SECTION _____

Negative Stain

1. Draw the cell shapes and arrangements you observed.

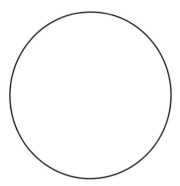

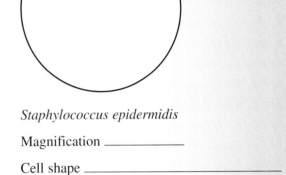

Bacillus cereus

Magnification _____

Cell shape _____

Cell arrangement _____

Staphylococcus epidermidis

Magnification _____

Cell shape _____

Cell arrangement _____

2. Answer the following questions in the space provided.

a. Explain why nigrosin, india ink, and congo red do not stain bacterial cells.

b. What are the advantages of negative stains?

Smear Preparation and the Simple Stain

Background

The use of a single basic stain to color bacterial cells is called a **simple stain**. Basic stains employed for this purpose include safranin, crystal violet, and methylene blue. These stains color the bacterial cells so that they are clearly visible with the microscope (figure 10.1). Since this procedure requires the heat-fixation of a smear prior to stain application, it does result in some cell shrinkage.

In this exercise, you will use a single basic stain to color the cells of several bacterial cultures to reveal their morphological characteristics.

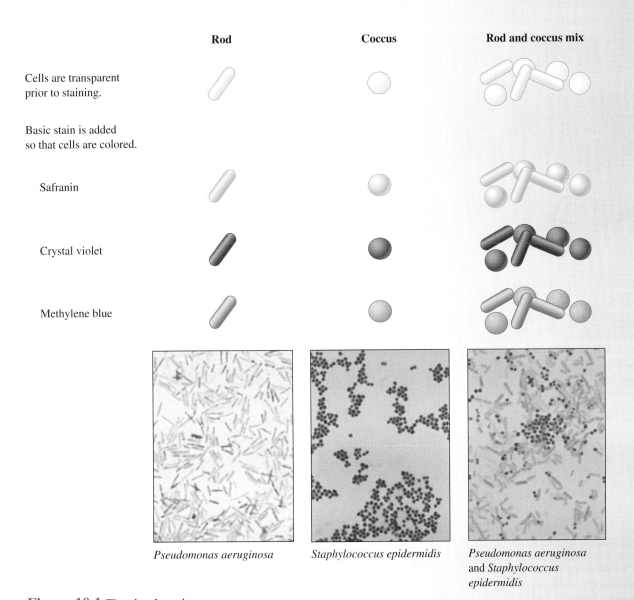

Figure 10.1 The simple stain.

Materials

Cultures (24–48 hour broth or agar)
  Pseudomonas aeruginosa (rod)
  Staphylococcus epidermidis (coccus)

Stains
  Crystal violet, methylene blue, or safranin

Equipment
  Light microscope

Miscellaneous supplies
  Bibulous paper
  Bunsen burner and striker
  Clothespin
  Disposable gloves (optional)
  Glass slides
  Immersion oil
  Inoculating loop or needle

Lens paper
Staining tray
Water bottle with tap water
Wax pencil

Procedure

Smear Preparation

1. Aseptically obtain a loopful of a broth culture of *Pseudomonas aeruginosa* by following the steps described in Exercise 9 (figure 9.2). If an agar culture is used instead of broth, follow the steps depicted in figure 10.2. *Note:* A loop or needle can be used to transfer from an agar culture; in either case, transfer only a pinhead amount of growth.

Figure 10.2 Bacterial smear preparation and the simple stain procedure.

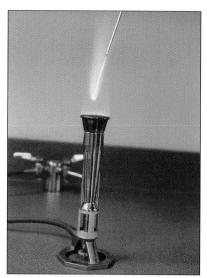

(a) Flame the loop (or needle) to red-hot to sterilize.

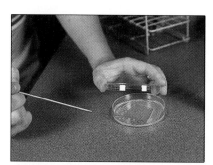

(b) Touch the loop (or needle) to an isolated colony to pick up a pinhead amount of growth.

(c) Transfer the growth to a drop of water on a slide and thoroughly mix to obtain a slightly milky color. This mixture must be air-dried and heat-fixed before staining.

(d) Cover the heat-fixed smear with stain and allow to sit for 60 seconds.

(e) After 60 seconds, wash off the stain with a water rinse. After drying, the stained smear is ready to observe.

2. Transfer the loopful of broth culture to a glass slide, and spread it out into a circle. If an agar culture is used instead of broth, mix the pinhead amount of culture into a drop of water on a glass slide as shown in figure 10.2c. Prepare a mixture that is only slightly milky in color. **Do not prepare a heavy suspension!**

3. Allow the slide to air-dry before heat-fixation. **Heat gently** by passing the slide over the flame several times. After heat-fixation, the slide is ready to stain.

4. Repeat steps 1–3 to prepare a smear of *Staphylococcus epidermidis* and to prepare a smear of a mixture of *Pseudomonas aeruginosa* and *Staphylococcus epidermidis*.

Simple Stain

1. Apply crystal violet, methylene blue, or safranin to the three smears, and let stand for 60 seconds (figure 10.2d). **Cover the entire smear with stain!**

2. After 60 seconds, gently wash off the stain with tap water (figure 10.2e). Blot the slide with bibulous paper, and examine using the oil-immersion objective.

LABORATORY REPORT

10

NAME _____ DATE _____

LAB SECTION _____

Smear Preparation and the Simple Stain

1. Draw your results from the simple stains.

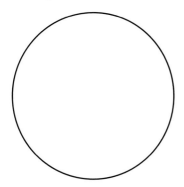

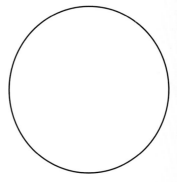

Pseudomonas aeruginosa

Magnification _____

Cell shape _____

Stain used _____

Color of cells _____

Staphylococcus epidermidis

Magnification _____

Cell shape _____

Cell arrangement _____

Stain used _____

Color of cells _____

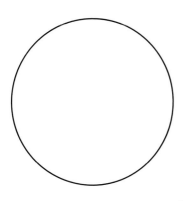

Pseudomonas aeruginosa and
Staphylococcus epidermidis mix

Magnification _____

Cell shapes _____

Cell arrangement _____

Stain used _____

Color of cells _____

2. Answer the following questions in the space provided.

    a. What is the purpose of heat-fixation? What happens if you heat-fix too much?

    b. Was your smear too thick when you viewed it (i.e., cells clumped too close together)? Why is a thick smear undesirable?

    c. Explain why methylene blue, crystal violet, and safranin stain differently from nigrosin, india ink, and congo red.

    d. What are the advantages and disadvantages of a simple stain?

    e. Although crystal violet and safranin can be used as simple stains, can you think of any reason it would be preferable to use methylene blue?

Gram Stain

Background

Most bacteria possess a cell wall that contains either a thick **peptidoglycan layer** or a thin peptidoglycan layer with an additional outer membrane composed of **lipopolysaccharide** (figure 11.1). This chemical difference in bacterial cell walls is identified with the Gram stain. The Gram stain is the stain most frequently used to identify unknown bacterial cultures, because it yields information on Gram reaction, cell size, cell shape, and cell arrangement.

During the Gram-staining procedure, all bacteria are stained purple by crystal violet, the primary stain. Bacterial cells that have a thick peptidoglycan layer retain the crystal violet during subsequent decolorization and counterstain steps. These bacteria appear purple when viewed with the microscope and are referred to as **Gram-positive** (figure 11.2). Bacterial cells that have a thin peptidoglycan layer and an added outer lipopolysaccharide layer lose the crystal violet during the decolorization step and take up the counterstain safranin. These bacteria appear red when viewed with the microscope and are referred to as **Gram-negative** (figure 11.2).

In this exercise, you will use the Gram stain on selected 18–24-hour bacterial cultures, as well as on a sample from your teeth or from yogurt.

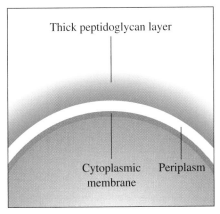

(a) Gram-positive

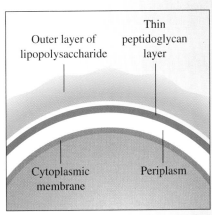

(b) Gram-negative

Figure 11.1 Differences in bacterial cell walls. (a) Gram-positive. (b) Gram-negative.

Materials

Cultures to select from (18–24-hour broth
or agar)
  Bacillus cereus (Gram-positive rod)
  Enterobacter aerogenes (Gram-negative rod)

Enterococcus faecalis (Gram-positive coccus)
Escherichia coli (Gram-negative rod)
Neisseria sicca (Gram-negative coccus)
Proteus vulgaris (Gram-negative rod)
Pseudomonas aeruginosa (Gram-negative rod)
Staphylococcus epidermidis (Gram-positive
  coccus)

| | Gram-positive | | Gram-negative | | Gram-negative rod and |
| | Rod | Coccus | Rod | Coccus | Gram-positive coccus mix |

Cells are transparent prior to staining.

Cells are colored purple by primary stain crystal violet and mordant Gram's iodine.

The decolorizing agent, ethyl alcohol, removes purple from Gram-negative cells; Gram-positive cells retain stain.

Gram-negative cells take up the counterstain, safranin, and are colored red; Gram-positive cells remain purple.

Bacillus cereus

Escherichia coli

E. coli and *S. epidermidis*

Staphylococcus epidermidis

Pseudomonas aeruginosa

Neisseria sicca

Enterococcus faecalis

Proteus vulgaris

Enterobacter aerogenes

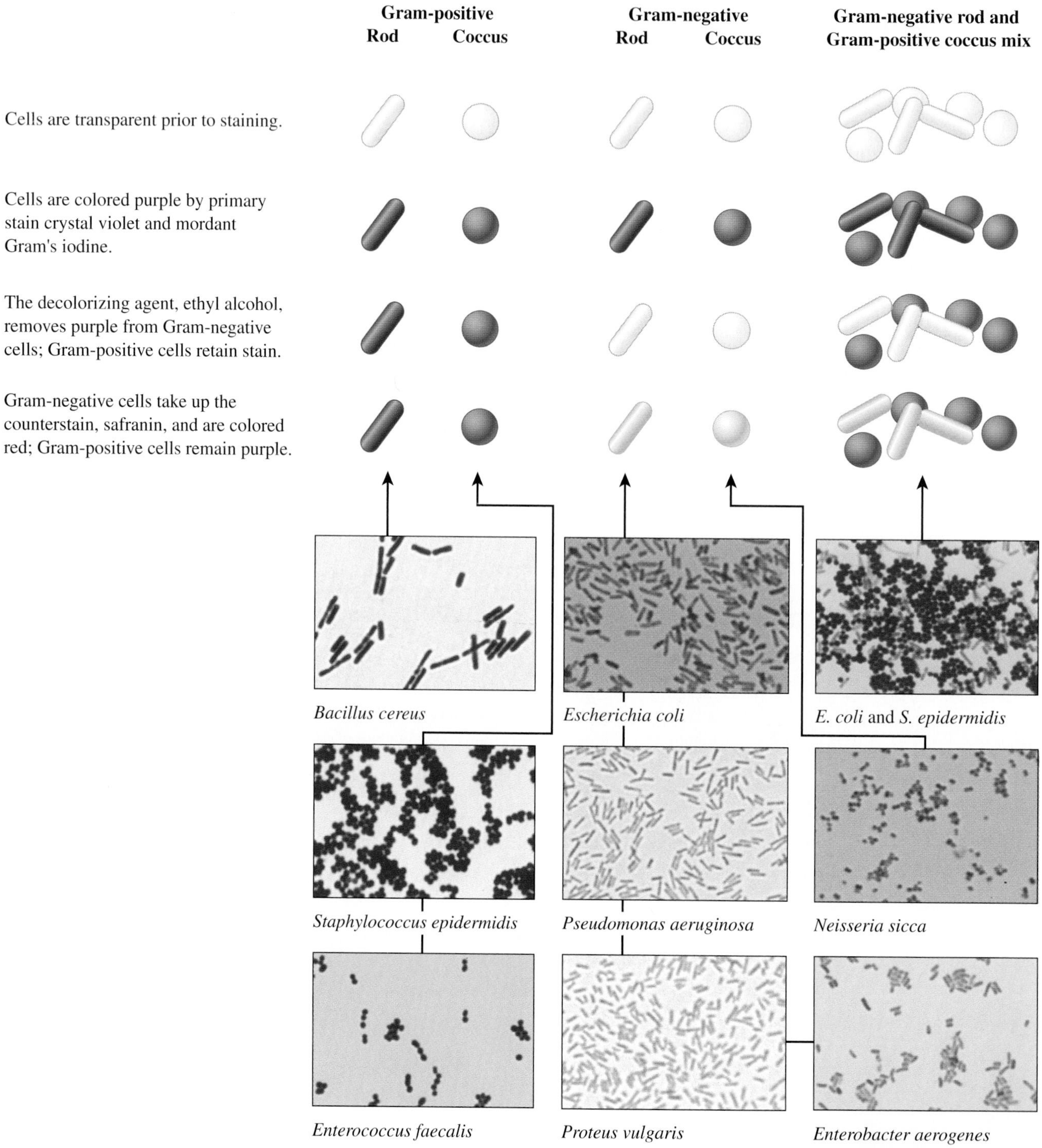

Figure 11.2 Gram stain results for Gram-positive and Gram-negative cells.

Stains
    Crystal violet
    Gram's iodine
    Ethanol (95%)
    Safranin

Equipment
    Light microscope

Miscellaneous supplies
    Bibulous paper
    Bunsen burner and striker
    Clothespin
    Disposable gloves (optional)
    Glass slides
    Immersion oil
    Inoculating loop or needle
    Lens paper
    Staining tray
    Toothpick
    Water bottle with tap water
    Wax pencil
    Yogurt

Procedure

Select four bacterial cultures from the materials list. Although a variety of bacteria can be used, one from each of the following categories is recommended: Gram-positive rod, Gram-positive coccus, Gram-negative rod, Gram-negative coccus, and a mixture of a Gram-positive coccus and a Gram-negative rod.

Smear Preparation

1. Following the steps outlined in Exercise 10 (figure 10.2), prepare smears of the four selected bacterial cultures and a smear of a mixture. **Wash your hands before proceeding.**

2. Obtain a clean toothpick from the container, and use it to prepare a smear of scrapings from your teeth, or from yogurt. To prepare a smear of teeth scrapings, use the end of a toothpick to pick material from between your teeth, and transfer it to a drop of water on a glass slide. Break up and mix the material into the drop as much as possible using the end of the toothpick. When finished, place the toothpick in a container of disinfectant. Allow this mixture to air-dry, and then heat-fix. To prepare a smear of yogurt, dip a toothpick in a container of yogurt, and transfer a small amount to a drop of water on a glass slide. Mix the yogurt into the drop using the toothpick. Allow this to air-dry before heat-fixation.

Gram Staining

1. Using the steps outlined in figure 11.3, Gram-stain all prepared smears. **Follow these steps exactly as outlined. Do not over-decolorize!** Tilt the slide, and drip alcohol onto the smear until it runs off clear. Stop decolorization at this point!

2. After Gram staining, examine all slides using the oil-immersion objective. *Note:* Avoid viewing areas of the slide where cells are clumped together. Only view areas where individual cells can be seen.

 **(a)** Apply crystal violet for 1 minute.

 **(b)** Rinse for 5 seconds with water.

 **(c)** Cover with Gram's iodine for 1 minute.

 **(d)** Rinse for 5 seconds with water.

 **(e)** Decolorize with 95% ethanol for 15–30 seconds.

 **(f)** Rinse for 5 seconds with water.

 **(g)** Counterstain with safranin for 1 minute.

 **(h)** Rinse for 5 seconds with water.

 **(i)** Blot dry with bibulous paper.

Figure 11.3 Gram-stain procedure.

LABORATORY REPORT

NAME _____ DATE _____

LAB SECTION _____

Gram Stain

1. Draw the results of your Gram stains.

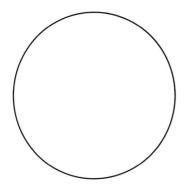

Gram-positive rod

Organism _____

Magnification _____

Cell shape _____

Cell arrangement _____

Cell color _____

Gram reaction _____

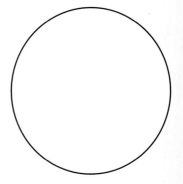

Gram-positive coccus

Organism _____

Magnification _____

Cell shape _____

Cell arrangement _____

Cell color _____

Gram reaction _____

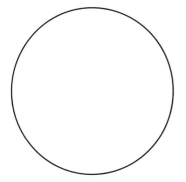

Gram-negative rod

Organism _____

Magnification _____

Cell shape _____

Cell arrangement _____

Cell color _____

Gram reaction _____

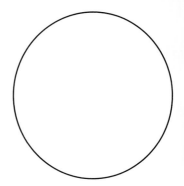

Gram-negative coccus

Organism _____

Magnification _____

Cell shape _____

Cell arrangement _____

Cell color _____

Gram reaction _____

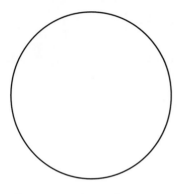

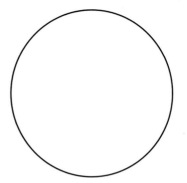

Gram-negative rod
and Gram-positive coccus mix

Organisms _____

Magnification _____

Cell shapes _____

Cell arrangements _____

Cell colors _____

Gram reactions _____

Teeth scrapings or yogurt

Sample _____

Magnification _____

Cell shapes _____

Cell arrangements _____

Cell colors _____

Gram reactions _____

2. Answer the following questions in the space provided.

a. Why are contrasting colors important in the Gram stain?

b. Explain why the alcohol decolorization step is so critical in the Gram stain.

c. Explain how a Gram stain differs from a:

(1) negative stain

(2) simple stain

d. Why is an 18–24-hour culture necessary for a Gram stain?

e. Name several pathogenic Gram-positive and Gram-negative bacteria and the diseases they cause.

Acid-fast Stain

Background

The acid-fast stain is used to distinguish certain bacteria that contain a high content of the lipid **mycolic acid** in their cell wall. This component makes the cell wall resistant to most stains, but heated carbolfuchsin will penetrate the cell wall, imparting a red color to cells that is not removed when the decolorizing agent, acid-alcohol, is added. Bacteria with this characteristic are referred to as **acid-fast** (figure 12.1). The majority of bacteria do not have as high a lipid content in their cell wall, so their cells lose the red color when acid-alcohol is added. They then take up the counterstain methylene blue. These bacteria are referred to as **non-acid-fast** (figure 12.1).

Several pathogenic bacteria can be distinguished by the acid-fast stain, including two species of **mycobacteria**—*Mycobacterium tuberculosis,* the causative agent of tuberculosis, and *Mycobacterium leprae*, the causative agent of leprosy. An acid-fast stain of sputum is important in the diagnosis of tuberculosis. In addition, certain pathogenic species of the actinomycete genus *Nocardia* are acid-fast, including *Nocardia asteroides*, a causative agent of nocardiosis. The oocysts of the sporozoan parasite *Cryptosporidium* are also acid-fast.

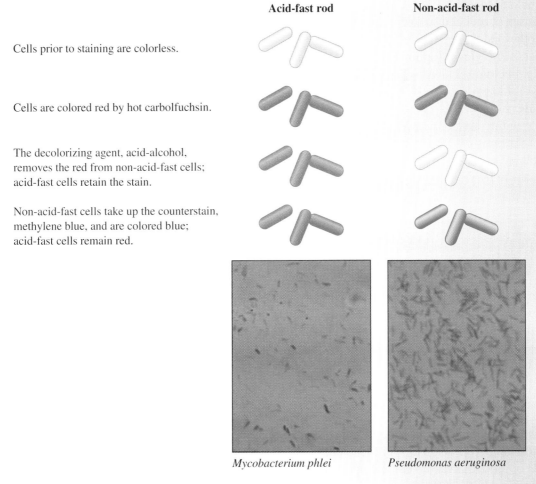

Cells prior to staining are colorless.

Cells are colored red by hot carbolfuchsin.

The decolorizing agent, acid-alcohol, removes the red from non-acid-fast cells; acid-fast cells retain the stain.

Non-acid-fast cells take up the counterstain, methylene blue, and are colored blue; acid-fast cells remain red.

Acid-fast rod

Non-acid-fast rod

Mycobacterium phlei

Pseudomonas aeruginosa

Figure 12.1 Acid-fast staining procedure.

In this exercise, you will use the Ziehl-Neelsen acid-fast stain method to demonstrate the acid-fast trait in *Mycobacterium phlei*.

Materials

Cultures (5–7-day agar)
Mycobacterium phlei (acid-fast rod)
Pseudomonas aeruginosa (non-acid-fast rod)

 All agents in red are BSL2 bacteria.

Stains
    Carbolfuchsin
    Acid-alcohol
    Methylene blue

Equipment
    Hot plate (optional, for heating carbolfuchsin)
    Light microscope

Miscellaneous supplies
    Bibulous paper
    Bunsen burner and striker
    Clothespin
    Disposable gloves (optional)
    Egg albumin solution
    Glass slides
    Immersion oil
    Inoculating loop or needle
    Lens paper
    Staining tray
    Water bottle with tap water
    Wax pencil

Procedure

Smear Preparation

1. Prepare a smear of *Mycobacterium phlei*, an acid-fast rod, and a smear of *Pseudomonas aeruginosa*, a non-acid-fast rod. Follow the steps outlined in Exercise 10 (see figure 10.2), with one exception: Mix the culture into a drop of egg albumin solution, instead of water. This solution will help acid-fast cells adhere to the glass slide. *Note:* If you have trouble transferring cells with a needle, use a loop instead. Since mycobacteria tend to clump together, use your inoculating needle or loop to break up cell clumps as much as possible into the drop.

Acid-fast Stain

1. After smear preparation, follow the steps of the Ziehl-Neelsen acid-fast staining procedure in figure 12.2. *Note:* The carbolfuchsin can be heated using either a hot plate (as depicted) or a Bunsen burner flame. In either case, **gently steam only; do not boil.** As the paper dries out, add more carbolfuchsin to keep the paper moist. After 5 minutes, remove the paper and continue the steps as outlined.

2. After staining, examine both slides using the oil-immersion objective.

(a) Apply carbolfuchsin to saturate paper, and heat for 5 minutes in an exhaust hood.

(b) Remove paper, cool, and rinse with water for 30 seconds.

(c) Decolorize with acid-alcohol until pink (10−30 seconds).

(d) Rinse with water for 5 seconds.

(e) Counterstain with methylene blue for about 2 minutes.

(f) Rinse with water for 30 seconds.

(g) Blot dry with bibulous paper.

Figure 12.2 Ziehl-Neelsen acid-fast staining procedure.

LABORATORY REPORT

NAME _____ DATE _____

LAB SECTION _____

Acid-fast Stain

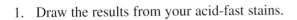

1. Draw the results from your acid-fast stains.

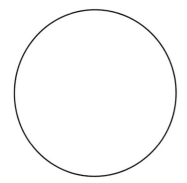
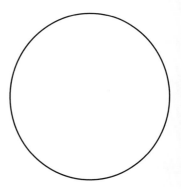

Mycobacterium phlei

Magnification _____

Cell shape _____

Cell color _____

Acid-fast? _____

Pseudomonas aeruginosa

Magnification _____

Cell shape _____

Cell color _____

Acid-fast? _____

2. Answer the following questions in the space provided.

   a. Explain these terms. Which one applies to species of *Mycobacterium* and *Nocardia?*

   (1) acid-fast

   (2) non-acid-fast

   b. Name several pathogenic acid-fast bacteria and the diseases they cause.

Spore Stain

Background

Some bacteria produce an internal structure known as an **endospore** during their life cycle (figure 13.1). This structure is produced by the vegetative cell by a process called **sporogenesis** and is released upon the death of the cell. The resulting **free spore** is a dormant structure that contains little water and carries out few chemical reactions. Its highly resistant nature is due to two factors: (1) a multilayered outer covering containing peptidoglycan; and (2) the presence of a protein-stabilizing molecule called **dipicolinic acid**. These components allow spores to survive adverse conditions that no other living thing could survive. When favorable conditions return, the bacterial spore undergoes **germination** to yield a vegetative cell (figure 13.1).

The detection of endospores is a useful characteristic in the identification of some bacteria, including species of *Bacillus* and *Clostridium*. Several species of these genera are pathogenic: *Bacillus anthracis* causes anthrax; *Clostridium botulinum* causes botulism; *Clostridium tetani* causes tetanus; and *Clostridium perfringens* causes gas gangrene.

When spores are detected in bacteria, their size, shape, and location are useful in identification. For example, *Bacillus cereus* and *Bacillus anthracis* produce an oval-shaped spore located in the center of the cell (central) (figure 13.2a). The spores of *Bacillus anthracis* are small enough that, when inhaled, they can enter the alveoli of the lungs, causing a disease known as inhalation anthrax. *Clostridium botulinum* produces oval-shaped spores located between the center and the end of the cell (subterminal) (figure 13.2b). *Clostridium tetani* produces spherical endospores at the end of the cell (terminal) (figure 13.2c). Terminal spores give this organism its characteristic "drumstick" appearance.

Due to their unique physical and chemical characteristics, endospores do not readily stain using ordinary staining procedures. However, basic stains easily color the vegetative cells that produce endospores. As a result, endospores can be seen in outline against the background of stained vegetative cells. For best observation and verification of their presence, spores should be stained using a special procedure called the **spore stain**.

In a spore stain, the cells are heated in the presence of malachite green. Heating drives the malachite green into the spore, where it is retained even during the water rinse step. When viewed with the microscope, spores are readily visible as green, oval-shaped or spherical objects within or outside of vegetative cells (figure 13.3). Water removes the malachite green from vegetative cells, allowing them to pick up the counterstain safranin and appear red. Non-spore-forming bacteria appear as red rods with no green, oval-shaped or spherical objects (figure 13.3).

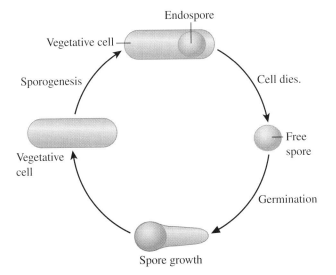

Figure 13.1 The life cycle of endospore-forming bacteria.

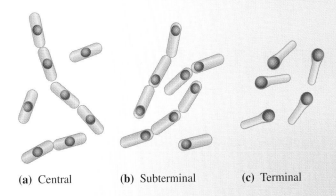

(a) Central   (b) Subterminal   (c) Terminal

Figure 13.2 Location of an endospore.

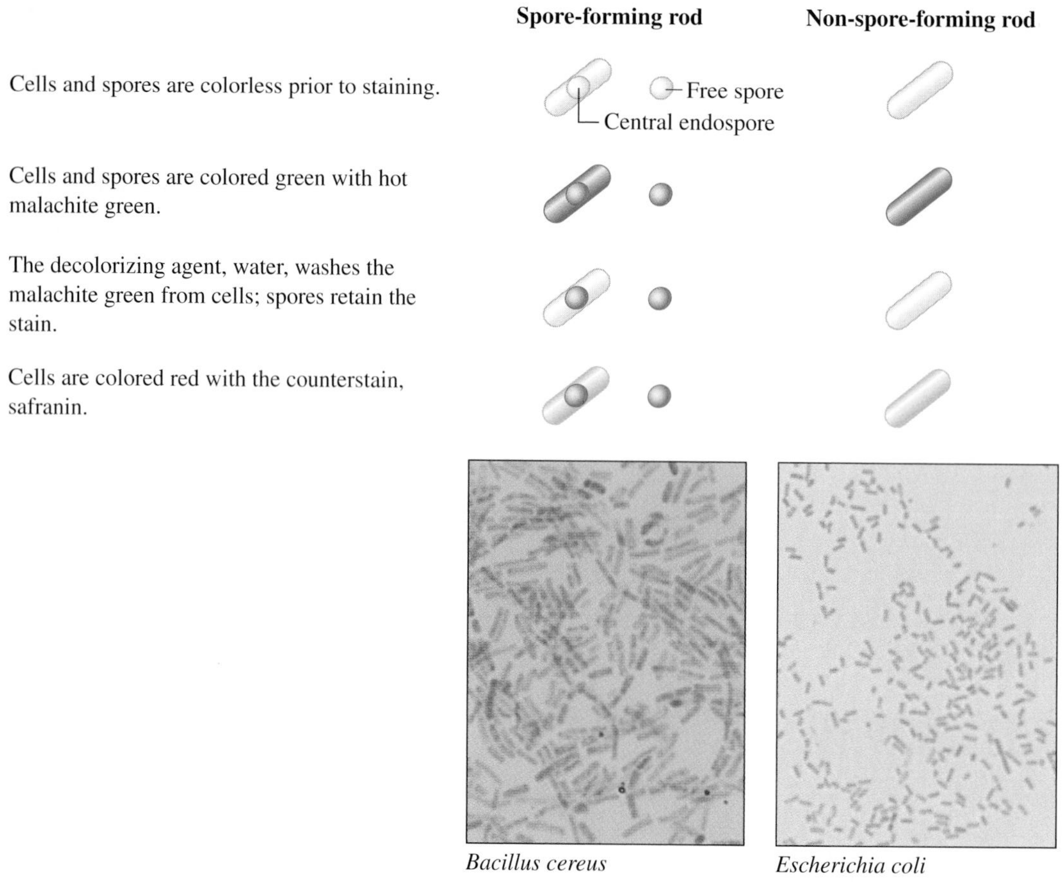

	Spore-forming rod	Non-spore-forming rod

Cells and spores are colorless prior to staining.

Free spore
Central endospore

Cells and spores are colored green with hot malachite green.

The decolorizing agent, water, washes the malachite green from cells; spores retain the stain.

Cells are colored red with the counterstain, safranin.

Bacillus cereus   *Escherichia coli*

Figure 13.3 The spore stain.

In this exercise, you will use the Schaeffer-Fulton endospore stain method to demonstrate the presence of spores in *Bacillus cereus*.

Materials

Cultures (4–5 days on nutrient agar)
   Bacillus cereus (spore-forming rod)
   Escherichia coli (non-spore-forming rod)

Stains
   Malachite green
   Safranin

Equipment
   Hot plate (optional, to heat malachite green)
   Light microscope

Miscellaneous supplies
   Bibulous paper
   Bunsen burner and striker
   Clothespin
   Disposable gloves (optional)
   Glass slides
   Immersion oil
   Inoculating loop or needle
   Lens paper
   Staining tray
   Water bath (optional, to heat malachite green)
   Water bottle with tap water
   Wax pencil

Procedure

Smear Preparation

1. Prepare a smear of *Bacillus cereus*, a spore-forming rod, and a smear of *Escherichia coli*, a non-spore-forming rod, following the steps outlined in Exercise 10 (see figure 10.2).

Spore Stain

1. After smear preparation, follow the steps of the Schaeffer-Fulton endospore stain method depicted in figure 13.4. *Note:* Heating the malachite green can be done using a water bath (as depicted), a hot plate, or a Bunsen burner flame. In either case, **gently steam only; do not boil**. As the paper dries out, add more malachite green to keep the paper moist. After 5 minutes, remove the paper and continue the steps as outlined.

2. After staining, examine both slides using the oil-immersion objective.

(a) Apply malachite green to saturate paper, and steam for 5 minutes.

(b) Remove paper, cool, and rinse with water for 30 seconds.

(c) Counterstain with safranin for 60–90 seconds.

(d) Rinse with water for 30 seconds.

(e) Blot dry with bibulous paper.

Figure 13.4 Schaeffer-Fulton endospore staining procedure.

LABORATORY REPORT

13

NAME _____ DATE _____

LAB SECTION _____

Spore Stain

1. Draw your results of the spore stains.

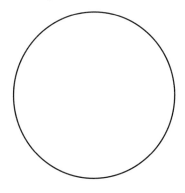

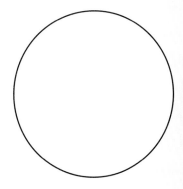

Bacillus cereus

Magnification _____

Cell shape _____

Vegetative cell color _____

Endospores? _____

    Color _____

    Location in cell _____

Free spores? _____

    Color _____

Escherichia coli

Magnification _____

Cell shape _____

Vegetative cell color _____

Endospores? _____

Free spores? _____

2. Answer the following questions in the space provided.

   a. Define these terms:

    (1) endospore

    (2) sporogenesis

    (3) germination

b. Explain why sporogenesis is not a form of bacterial reproduction.

c. How do bacterial endospores differ from mold asexual spores (conidia)?

d. Why is a 4–5-day culture of *Bacillus cereus* required for this exercise instead of a 1–2-day culture?

e. Why is heat applied to the malachite green in the spore stain? What function does water serve in this method?

f. Name several spore-forming pathogens and the diseases they cause.

3. Answer the following questions based on these photographs:

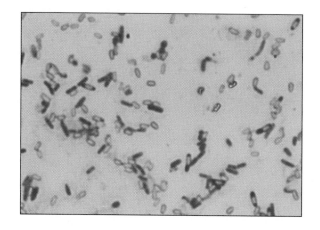
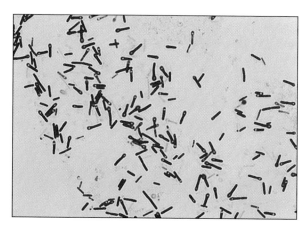

a. What are the clear ovals? _____

   Why are they not stained? _____

   What is stained? _____

b. What is your tentative identification of this spore-former?

   _____

Capsule and Flagella Stains

Background

Some bacteria have cell structures external to the cell wall that are visible with the light microscope after special staining. One of these structures is a **capsule**, an extracellular layer surrounding the cell wall that is composed of polysaccharides and polypeptides. Although a capsule is resistant to staining, it can be revealed by using a combination of acidic and basic stains. The acidic stain colors the background, while the basic stain colors the cell. The capsule appears as a clear halo around the cell. Non-capsule-forming bacteria do not have a halo around the cell (figure 14.1).

Several clinically important bacteria form capsules, including *Klebsiella pneumoniae* and *Streptococcus pneumoniae*, both causes of bacterial pneumonia. In these and other bacteria, the capsule is considered a virulence factor, since it protects the cell from phagocytosis by white blood cells.

A second external cell structure that is visible after staining is the **flagellum**, a long, whiplike structure composed of protein and used by bacteria for

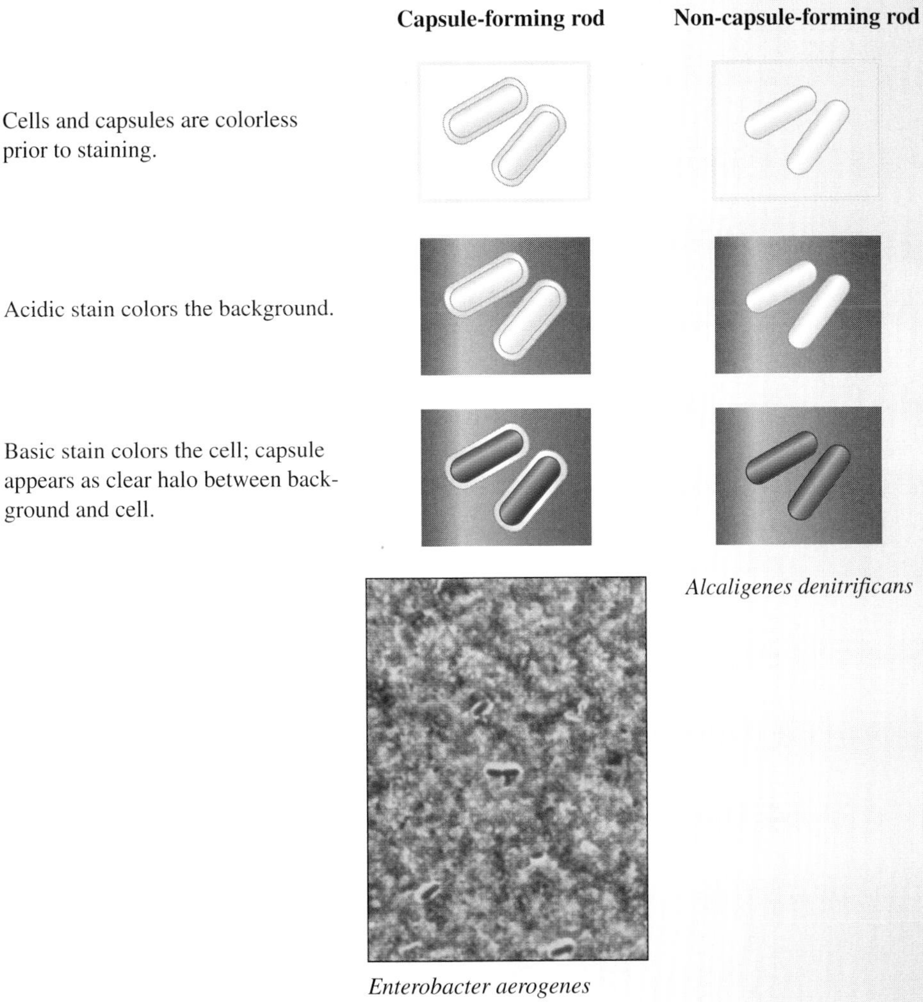

Capsule-forming rod   **Non-capsule-forming rod**

Cells and capsules are colorless prior to staining.

Acidic stain colors the background.

Basic stain colors the cell; capsule appears as clear halo between background and cell.

Alcaligenes denitrificans

Enterobacter aerogenes

Figure 14.1 Capsule stain.

motility. Bacteria that possess a flagellum (one) or flagella (two or more) are referred to as **motile**, while bacteria that lack this structure are referred to as **nonmotile**. Although motility test agar can determine if bacteria are motile or nonmotile (see Exercise 18), it does not provide information about the number or arrangement of flagella. Only a flagella stain can provide this information. In a flagella stain, the stain clumps around the surface of the flagella, widening their diameter so that they can be seen with a light microscope. Microscopic observation reveals several flagella arrangements in bacteria: **monotrichous, amphitrichous, lophotrichous,** and **peritrichous** (figure 14.2). These arrangements are useful in bacterial identification.

In this exercise, you will prepare a capsule stain of an encapsulated and nonencapsulated culture. You will not prepare flagella stains, since flagella are very fragile and easily break off bacterial cells; but you will examine several prepared slides of flagella stains.

Materials

Cultures (18–24-hour broth)
  Enterobacter aerogenes (encapsulated rod)
  Alcaligenes denitrificans (nonencapsulated rod)

Stains
  Acidic: india ink
  Basic: crystal violet

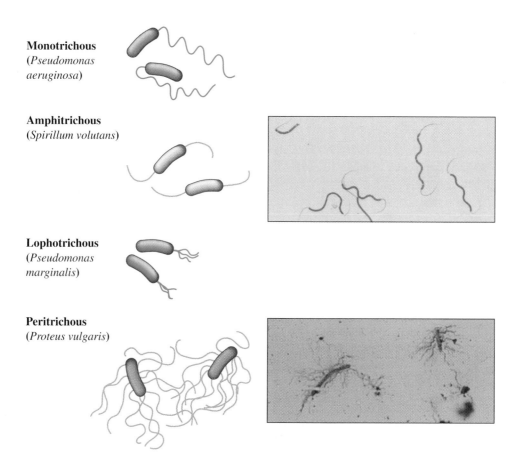

Figure 14.2 Flagella arrangements in bacteria.

Prepared slides
    Proteus vulgaris (peritrichous flagella)
    Spirillum volutans (amphitrichous flagella)

Equipment
    Light microscope

Miscellaneous supplies
    Bibulous paper
    Bunsen burner and striker
    Clothespin
    Disposable gloves (optional)
    Glass slides
    Immersion oil
    Inoculating loop
    Lens paper
    Staining tray
    Water bottle with tap water
    Wax pencil

Procedure

Capsule Stain

1. Study the steps for preparing a capsule stain shown in figure 14.3. Then carefully follow this procedure as you prepare a capsule stain of two cultures: *Enterobacter aerogenes*, an encapsulated rod, and *Alcaligenes denitrificans*, a nonencapsulated rod. **Gently heat only; gently rinse with water.**

2. When finished staining, examine both slides using the oil-immersion objective.

Flagella Stain

Examine the prepared slides of flagella stains using the oil-immersion objective. Note the number and arrangement of flagella.

(a) Two loopfuls of the organism are mixed in a small drop of india ink.

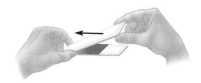

(b) The ink suspension of bacteria is spread over the slide and air-dried.

(c) The slide is *gently* heat-dried to fix the organisms to the slide.

(d) Smear is stained with crystal violet for 1 minute.

(e) Crystal violet is *gently* washed off with water.

(f) Slide is blotted dry with bibulous paper, and examined with oil-immersion objective.

Figure 14.3 Procedure for demonstration of capsule.

LABORATORY REPORT

14

NAME _____ DATE _____

LAB SECTION _____

Capsule and Flagella Stains

1. Draw the results of your capsule stains.

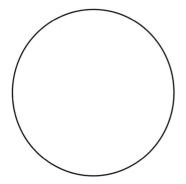

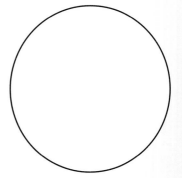

Enterobacter aerogenes

Magnification _____

Color of background _____

Color of cells _____

Cell shape _____

Capsule? _____

    Color _____

Alcaligenes denitrificans

Magnification _____

Color of background _____

Color of cells _____

Cell shape _____

Capsule? _____

2. Draw the prepared slides of flagella you examined.

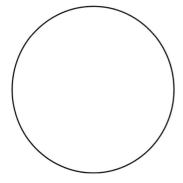

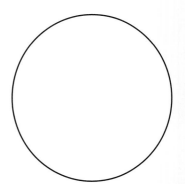

Spirillum volutans

Magnification _____

Cell shape _____

Flagella present? _____

    Arrangement _____

Proteus vulgaris

Magnification _____

Cell shape _____

Flagella present? _____

    Arrangement _____

3. Answer the following questions in the space provided.

    a. What are bacterial capsules? How do capsules play a role in the establishment of disease?

    b. Why must a combination of basic and acidic stains be used to reveal a capsule?

    c. Name several capsule-forming bacteria and the diseases they cause.

    d. Why are bacterial flagella visible with a light microscope only after a flagella stain?

    e. Motility provided by bacterial flagella can be observed in wet mount preparations. What additional information does a flagella stain provide? How can this information be useful?

4. Answer the following questions based on these photographs:

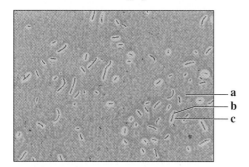

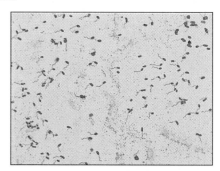

Is this organism encapsulated? _____

Identify (a) _____

Identify (b) _____

Identify (c) _____

Is this organism flagellated? _____

If yes, what is the arrangement? _____

Name one bacterium with this type of arrangement?

_____

The Staining Characterization of a Bacterial Unknown

Background

This exercise will allow you to apply what you have learned in previous exercises to the identification of a bacterial unknown. First, you will select or have assigned to you one of the bacterial cultures from the materials list, but you will not know which one. You will then use the information provided in figure 15.1 and table 15.1 to guide you through the characterization of your unknown. Perform only those staining procedures required as you work your way through the scheme. When you are finished, you should be able to correctly identify your bacterial unknown based on its staining characteristics.

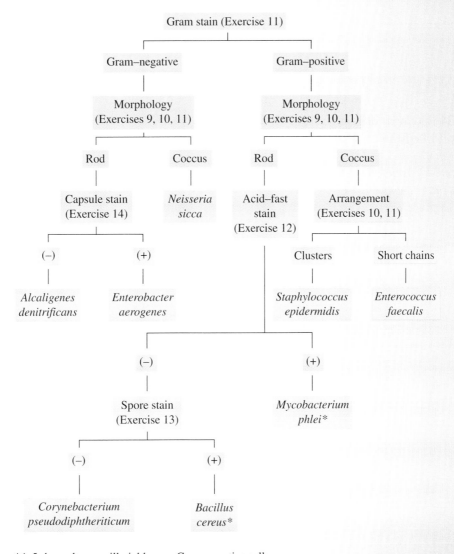

*4–5-day cultures will yield some Gram-negative cells.

Figure 15.1 Identification scheme for eight bacterial unknowns. You can use this scheme when identifying your staining unknown.

Table 15.1 Cell Size, Shape, and Arrangement of Bacterial Staining Unknowns

Bacterial culture	Cell size (μ)	Cell shape	Cell arrangement
Alcaligenes denitrificans	0.5 × 1–2	Rod	Single cells or pairs
Bacillus cereus	1 × 3–5	Rod	Streptobacilli
Corynebacterium pseudodiphtheriticum	0.5 × 1–2	Rod	Single cells or V-shapes
Enterobacter aerogenes	0.5 × 1–2	Rod	Single cells or pairs
Enterococcus faecalis	0.5–1	Coccus	Single cells, pairs, or short streptococci
Mycobacterium phlei	0.2 × 1–2	Rod	Single cells or cords
Neisseria sicca	0.5–1	Coccus	Single cells or diplococci
Staphylococcus epidermidis	0.5–1.5	Coccus	Single cells, pairs, or staphylococci

Materials

Cultures (18–24-hour agar or broth)
  Alcaligenes denitrificans
  Bacillus cereus
  Corynebacterium pseudodiphtheriticum
  Enterobacter aerogenes
  Enterococcus faecalis
  Mycobacterium phlei
  Neisseria sicca
  Staphylococcus epidermidis

 All agents in red are BSL2 bacteria.

Stains
  Gram stain
    Crystal violet
    Gram's iodine
    Ethanol (95%)
    Safranin

Acid-fast stain
  Carbolfuchsin
  Acid-alcohol
  Methylene blue

Spore stain
  Malachite green
  Safranin

Capsule stain
  Acidic stain: india ink
  Basic stain: crystal violet

Equipment
  Hot plate (optional)
  Light microscope

Miscellaneous supplies
  Bibulous paper
  Bunsen burner and striker
  Clothespin
  Disposable gloves (optional)
  Egg albumin solution
  Glass slides
  Immersion oil
  Inoculating loop or needle
  Lens paper
  Ocular micrometer
  Stage micrometer slide
  Staining tray
  Wash bottle with tap water
  Wax pencil

Procedure

1. You will select or have assigned to you an unknown from the materials list. Record your unknown number in the laboratory report.

2. Examine the information provided in figure 15.1 and table 15.1. Notice that the Gram stain must be done first to determine Gram reaction and cell morphology (figure 15.1). A negative stain or simple stain may be used in conjunction with a Gram stain to verify cell size, shape, and arrangement (table 15.1). Notice that Gram-positive and Gram-negative cocci will require no additional staining, but Gram-positive and Gram-negative rods will require one or more additional stains for identification. So, whatever your Gram stain results, continue to follow the identification scheme downward until you identify your unknown based on staining characteristics.

3. As you perform whatever stains are required, record your results in the laboratory report.

LABORATORY REPORT

15

NAME _____ DATE _____

LAB SECTION _____

The Staining Characterization of a Bacterial Unknown

Unknown no. _____

1. Follow the information provided in figure 15.1 and table 15.1 to identify your staining unknown. Perform only the stains required to identify your unknown.

2. Required staining results:

Gram stain
(See Exercise 11.)

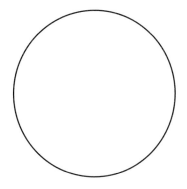

Acid-fast stain
(See Exercise 12.)

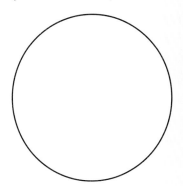

Magnification _____

Cell size _____

Cell shape _____

Cell arrangement _____

Cell color _____

Gram reaction _____

Magnification _____

Cell shape _____

Cell arrangement _____

Cell color _____

Acid-fast? _____

Spore stain
(See Exercise 13.)

Capsule stain
(See Exercise 14.)

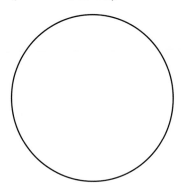

Magnification _____

Cell shape _____

Cell arrangement _____

Vegetative cell color _____

Endospores? _____

If yes, color? _____

If yes, location in cell _____

Free spores? _____

If yes, spore color?_____

Magnification _____

Color of background _____

Color of cells _____

Cell shape _____

Capsule? _____

If yes, capsule color? _____

3. Summary of the staining characteristics of my unknown:

Unknown no.	Cell shape	Cell arrangement	Gram reaction	Acid-fast?

Spores?	Capsules?

4. After examining the information provided in figure 15.1 and table 15.1 and recording the results of required stains, I conclude that my staining unknown is _____.

5. Based on the information provided in figure 15.1 and table 15.1, fill in the following table for the eight unknown cultures.

Unknown culture	Cell shape	Cell arrangement	Gram reaction	Acid-fast	Spores	Capsules
Alcaligenes denitrificans						
Bacillus cereus						
Corynebacterium pseudodiphtheriticum						
Enterobacter aerogenes						
Enterococcus faecalis						
Mycobacterium phlei						
Neisseria sicca						
Staphylococcus epidermidis						

Bacterial Cultivation

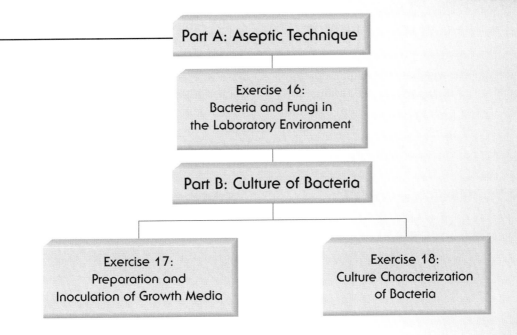

Part A: Aseptic Technique

Exercise 16:
Bacteria and Fungi in
the Laboratory Environment

Part B: Culture of Bacteria

Exercise 17:
Preparation and
Inoculation of Growth Media

Exercise 18:
Culture Characterization
of Bacteria

EXERCISE

16

Bacteria and Fungi in the Laboratory Environment: The Necessity of Aseptic Technique

Background

Bacteria and Fungi in the Laboratory Environment

Bacteria and fungi occur widely in the natural environment in association with air, water, soil, plants, and animals. These microorganisms find their way into our homes, offices, and buildings in a variety of ways: (1) through open doors and windows; (2) on the bottoms of shoes; (3) on the surfaces of plants, pets, and food; and (4) on the surfaces of our hands and clothes.

Bacteria and fungi also find their way into our laboratory environment, where they can be found in the air and on countertops (figure 16.1). We must be aware of these microorganisms in the laboratory environment when working with laboratory cultures.

To demonstrate their presence, you will use two types of media to culture bacteria and fungi from the laboratory environment: **nutrient agar** and **Sabouraud**

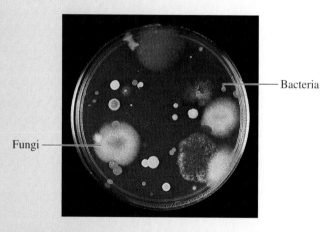

Bacteria

Fungi

(a) Bacteria and fungi from laboratory air.

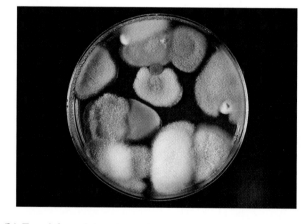

(b) Fungi from laboratory air.

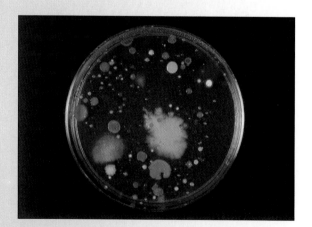

(c) Bacteria and fungi from laboratory countertop.

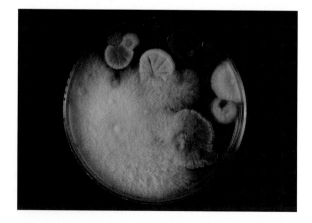

(d) Fungi from laboratory countertop.

Figure 16.1 Bacteria and fungi from the laboratory environment.

Table 16.1 Components of Nutrient Agar and Sabouraud Dextrose Agar			
Nutrient agar* **(bacteria)**		**Sabouraud dextrose** **agar (fungi)**	
Peptone	5 g	Peptone	10 g
Beef extract	3 g	Dextrose	40 g
Agar	15 g	Agar	15 g
Distilled water	1,000 ml	Distilled water	1,000 ml
Final pH =	6.8	Final pH =	5.6

Source: *The Difco Manual.* Eleventh Edition. Difco Laboratories.
*Nutrient broth has the same formula, but does not contain agar.

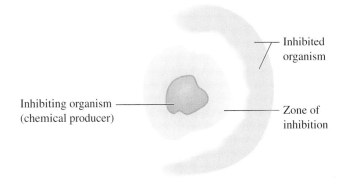

Figure 16.2 Evidence of inhibition of one microbe by another. The inhibiting organism is producing a chemical that is active against the other organism.

dextrose agar (table 16.1). Nutrient agar contains organic compounds, which support the growth of a wide variety of bacteria, and agar as a solidifying agent. The final pH of the medium is 6.8. Sabouraud dextrose agar also contains organic compounds and agar, but the high dextrose content (4%) and low pH (5.6) favor the growth of fungi over bacteria. The medium can be made even more selective for fungi through the addition of an antibiotic, such as chloramphenicol. Together, these two media will demonstrate the number and variety of bacteria and fungi in our laboratory environment.

Excluding Environmental Contaminants from Laboratory Cultures

Once your examination of culture media has revealed the existence of bacteria and fungi in our laboratory environment, you will be asked to consider how this relates to working with pure cultures in the laboratory. For example, can these environmental bacteria and fungi contaminate our laboratory cultures? Can their entry be prevented by using certain techniques designed to exclude them? If such techniques exist, what are they?

Searching for Examples of Antibiosis

The primary focus of this exercise is to demonstrate bacteria and fungi in the laboratory environment and to consider techniques to exclude them from cultures. A secondary focus is to find an example of **antibiosis** on the media you inoculate with samples from the laboratory environment. What is antibiosis? When envi-

ronmental microorganisms grow in close proximity to one another, as occurs naturally in soil or unnaturally in a culture medium, one microbe may produce a chemical substance that inhibits the growth of another microbe nearby. This phenomenon is called antibiosis.

Antibiosis is identified in culture media by a **zone of inhibition** around the chemical-producing organism (figure 16.2). Examples of antibiosis in culture media are not common, since the odds are low that you will inoculate in close proximity an organism that produces a chemical substance inhibitory to another. You and other laboratory students may collectively find only one example on all your plates, but finding that one example is the objective. When you see this example, you will understand what Alexander Fleming saw in 1928 when he examined a plate in his laboratory. He found that the growth of *Staphylococcus aureus* was inhibited by a mold. The mold was identified as *Penicillium,* and the chemical substance it produced was later isolated and named penicillin. Penicillin proved to be effective in treating infections caused by *Staphylococcus aureus* in the human body. It became the first **antibiotic**, an antimicrobial chemical agent of microbial origin put into the human body to treat disease. Since the introduction of penicillin, many other antibiotics of microbial origin have been discovered.

Streptomyces and *Bacillus*, Examples of Antibiotic-Producers

Streptomyces is a genus of bacteria that is common in soil. These bacteria are Gram-positive and produce rods in branching filaments similar to those of fungi, but

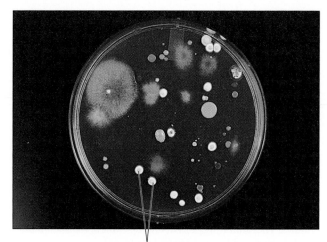

Streptomyces colonies

Figure 16.3 Swab results from the laboratory floor. A number of white, powdery colonies of *Streptomyces* are present.

the filaments have a much smaller diameter than those of fungi. *Streptomyces* produces a small, white, powdery colony on culture media (figure 16.3) and gives off an "earthy," soil-like odor.

Bacillus is a genus of bacteria also common in soil. This organism is a Gram-positive, endospore-forming rod. *Bacillus* generally produces a large, flat colony that is typically white or cream-colored.

Streptomyces and *Bacillus* are both sources of useful antibiotics. Species of *Streptomyces* are the source of more than half of all antibiotics effective against bacteria, including streptomycin, tetracycline, and chloramphenicol. They are also the source of the polyenes, such as amphotericin B and nystatin, effective against fungi. Species of *Bacillus* are the source of antibiotics such as bacitracin, effective against Gram-positive bacteria, and polymyxin B, effective against Gram-negative bacteria.

A third focus of this exercise is to find one or both of these common soil bacteria in the laboratory environment. One of these bacteria may provide the example of antibiosis in your culture media.

Materials

Media
  4 nutrient agar plates
  4 Sabouraud dextrose agar plates
  3 nutrient broth (or water) tubes, sterile

Stains
  Gram stain
    Crystal violet
    Gram's iodine
    Ethanol (95%)
    Safranin

Equipment
  Dissecting microscope
  Incubator (set at 35°C)
  Light microscope

Miscellaneous supplies
  Bibulous paper
  Bunsen burner and striker
  Clothespin
  Cotton-tipped swabs, sterile (3)
  Disposable gloves (optional)
  Glass slides
  Immersion oil
  Inoculating needle
  Lens paper
  Staining tray
  Wash bottle with tap water
  Wax pencil

Procedure

First Session: Inoculation of Nutrient Agar and Sabouraud Dextrose Agar Plates

1. Remove the lids from a nutrient agar plate and a Sabouraud dextrose agar plate. Leave these two plates open to the laboratory air for 30–60 minutes.

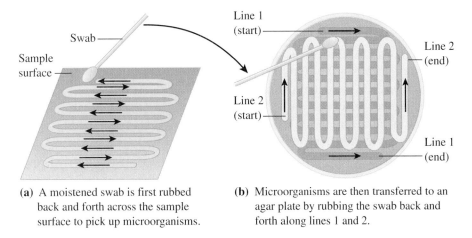

(a) A moistened swab is first rubbed back and forth across the sample surface to pick up microorganisms.

(b) Microorganisms are then transferred to an agar plate by rubbing the swab back and forth along lines 1 and 2.

Figure 16.4 Swab inoculation of an agar plate.

2. Dip the cotton-tipped end of a sterile swab into a tube of sterile water or nutrient broth. Blot the excess liquid against the test tube wall. Use the wetted end to rub back and forth across the countertop of your work area; then inoculate a nutrient agar plate (figure 16.4). Repeat this process to inoculate a Sabouraud dextrose agar plate.

3. Wet a second sterile, cotton-tipped swab with sterile water or broth, and use it to rub back and forth across the floor below your work area. Inoculate a second nutrient agar plate as before. Repeat to inoculate a second Sabouraud dextrose agar plate.

4. Wet a third sterile, cotton-tipped swab with sterile water or broth, and use it to rub back and forth across the skin on the inside of your left hand. Inoculate a third nutrient agar plate as before, and repeat to inoculate a third Sabouraud dextrose agar plate.

5. Label the plates, and incubate the air, countertop, and floor plates at room temperature (22°C). Place the skin plates in an incubator set at 35°C. Incubate all plates at least 3–4 days before examining.

Second Session: Examination of Nutrient Agar and Sabouraud Dextrose Agar Plates

Number and Variety of Bacteria and Fungi in the Laboratory Environment

1. After incubation, examine all plates for growth. Sketch a typical nutrient agar plate and Sabouraud dextrose agar plate in the laboratory report.

2. Record the number and variety of bacteria and fungi on your plates in the table of your laboratory report. *Note:* When determining number, count only bacterial colonies on nutrient agar plates and fungal colonies on Sabouraud dextrose agar plates. Bacterial colonies are smooth and round, while fungal colonies are large and cottony in appearance (see figure 16.1a). *Note:* When determining variety, look for bacterial and fungal colonies that are different in appearance. Colonies that look different (based on size, shape, margin, texture, elevation, pigmentation, etc.) represent different types. **The use of a dissecting microscope may help you count and differentiate colonies.**

Examples of Antibiosis on Plates

1. After completing the drawings and table and considering the implications of these results, go back through your plates searching for examples of antibiosis. Refer to figure 16.2 to determine if a zone of inhibition is present on any of your plates. If you find none, examine the plates of other students. Generally, at least one example can be found.

2. In the laboratory report, draw the example of antibiosis you see. If time permits, Gram-stain the two organisms in order to get an idea of what antibiotic may be involved.

Presence of *Streptomyces* and *Bacillus*

1. Examine your plates for signs of *Streptomyces* and *Bacillus*. Their colony characteristics were described previously in the "Background" section of this exercise.

2. If you find suspect colonies, do a Gram stain to verify your identification. Also, do a spore stain on the suspect *Bacillus* colony if it turns out to be a Gram-positive rod. If you have isolated species of one or both of these bacteria, you have isolated important antibiotic-producers. Did one of these bacteria provide your example of antibiosis?

LABORATORY REPORT

NAME _____ DATE _____

LAB SECTION _____

Bacteria and Fungi in the Laboratory Environment: The Necessity of Aseptic Technique

1. a. Draw a typical nutrient agar plate and Sabouraud dextrose agar plate inoculated with a laboratory sample.

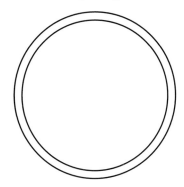

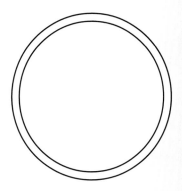

Nutrient agar plate (bacteria)

Sample _____

Total bacterial colonies _____

Total colony types _____

Sabouraud dextrose agar plate (fungi)

Sample _____

Total fungal colonies _____

Total colony types _____

   b. Record your results for all plates in the following table.

Laboratory sample	Nutrient agar (bacteria)		Sabouraud dextrose agar (fungi)	
	Total colonies	*Colony types*	*Total colonies*	*Colony types*
Air				
Countertop				
Floor				
Skin				

Which sample had the highest number of bacteria? _____Why? _____

_____

   c. Based on your results, does the laboratory environment contain a large number and variety of bacteria and fungi?

d. Describe several techniques you might use to keep these environmental bacteria and fungi from contaminating your laboratory cultures.

2. a. Did you see any zones of inhibition on your plates? (yes or no) _____

   Any zones on other students' plates? (yes or no) _____

   b. If yes, draw a representative result indicating a zone of inhibition. Draw only the region of the interacting organisms. Label the chemical-producing organism, the zone of inhibition, and the inhibited organism.

   c. Explain how your result is similar to that observed by Alexander Fleming in 1928.

   d. Why was Fleming's observation historically important?

   e. Gram-stain results: Antibiotic-producer in the drawing in (b):

                    Gram reaction _____

                    Cell shape _____

                    Bacteria or fungi? _____

          Inhibited organism in the drawing in (b):

                    Gram reaction _____

                    Cell shape _____

                    Bacteria or fungi? _____

   Do these results give you any clues as to what antibiotic is being produced? If so, describe the possibilities.

3. a. Did you or another student have an isolate from the countertop or floor with the following characteristics:

   Small, white, powdery colony? (yes or no) _____

   Gram-positive rods in branching filaments? (yes or no) _____

   If you answered yes on both lines, you may have isolated *Streptomyces*, a common bacterium in soil.

   Can you explain how this organism gets into the lab? _____

   _____

   What is the medical significance of this organism? _____

   _____

   b. Did you or another student have an isolate from the countertop or floor with the following characteristics:

   Large, flat colony, white or cream-colored? (yes or no) _____

   Gram-positive endospore-forming rod? (yes or no) _____

   If you answered yes on both lines, you may have isolated *Bacillus*, a common bacterium in soil.

   Can you explain how this organism gets into the lab? _____

   _____

   What is the medical significance of this organism? _____

   _____

4. Answer the following questions based on these photographs:

   a. Bacteria or fungi?

      (1) _____

      (2) _____

   b. Type of medium? _____

      How do you know?

      _____

   c. Bacteria from air or skin?

      _____

      How do you know?

      _____

Gram-negative rod

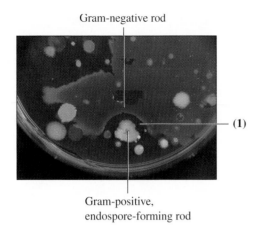

Gram-positive,
endospore-forming rod

(1)

Fungi

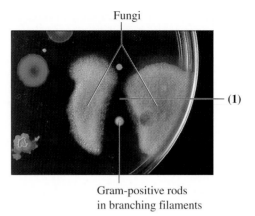

(1)

Gram-positive rods
in branching filaments

d. Name area of no growth (1) _____

   Given these results, what group of antibiotics may

   be indicated? _____

e. Name area of no growth (1) _____

   Given these results, what group of antibiotics may

   be indicated? _____

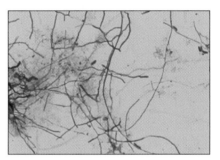

f. These two photographs indicate

   what type of bacterium? _____

   Where does this organism occur naturally?

   _____

   What is the medical significance

   of this organism? _____

17

Preparation and Inoculation of Growth Media

Background

Media Preparation

The cultivation of bacteria (i.e., their growth on a nutrient medium) is necessary for subsequent isolation and identification. A **complex medium**, one that contains an array of organic nutrients, can grow a variety of bacteria. Media of this type include tryptic soy broth and tryptic soy agar (table 17.1). In this exercise, you will cultivate bacteria using different forms of these media, including **broth tubes, agar slants, agar deeps,** and **agar plates** (figure 17.1).

Media Sterilization

Before use, media must be sterilized in an **autoclave** (figure 17.2). **Sterilization** is a process that destroys all microbes in the medium. If autoclaving is not done, these microbes will contaminate the culture you introduce into the medium.

Media Inoculation

After media preparation and sterilization, a culture is inoculated (introduced) into each medium. Media **inoculation** can be done using a variety of sterile instruments, such as a loop, needle, swab, or pipette (figure 17.3). In all cases, care must be taken to avoid introducing environmental bacteria and fungi into the medium with the culture. To prevent contamination of your media, examine and carefully follow the **aseptic techniques** outlined in table 17.2. These procedures must become a standard part of your laboratory technique when working with pure cultures.

Table 17.1 Composition of Tryptic Soy Broth and Tryptic Soy Agar			
Tryptic soy broth		**Tryptic soy agar**[*]	
Tryptone	17 g	Tryptone	15 g
Soytone	3 g	Soytone	5 g
Dextrose	2.5 g	Sodium chloride	5 g
Sodium chloride	5 g	Agar	15 g
Dipotassium phosphate	2.5 g	Distilled water	1,000 ml
Distilled water	1,000 ml		
Final pH =	7.3	Final pH =	7.3

Source: *The Difco Manual.* Eleventh Edition. Difco Laboratories.
*The addition of 5% sheep blood makes blood agar, a medium used to cultivate fastidious bacteria and to determine hemolytic reactions.

Figure 17.1 Different forms of culture media. From left to right: broth tube, agar slant (front view), agar slant (side view), agar deep, and agar plate (top and side views).

Figure 17.2 An autoclave, used to sterilize media.

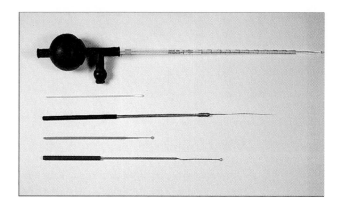

Figure 17.3 Instruments used to inoculate media. From bottom to top: standard loop, calibrated loop (disposable), needle, cotton-tipped swab, and pipette with safety bulb.

Media Incubation and Examination

Inoculated media must be incubated to allow time for bacterial growth. After incubation, the microbial growth in tubes and plates is visible and can be examined. During examination, look for contamination by unwanted bacteria and fungi. The presence of more than an occasional contaminant suggests the need for better aseptic technique.

Materials

Culture (24-hour broth)
   Escherichia coli

Media
   Tryptic soy agar
   Tryptic soy broth

Equipment
   Autoclave
   Balance
   Hot plate
   Incubator (set at 35°C)

Miscellaneous supplies
   Aluminum foil
   Bunsen burner and striker
   Distilled water
   Erlenmeyer flask, 250 ml
   Graduated cylinder, 100 ml
   Immersion oil
   Inoculating loop and needle
   Petri dishes, sterile (3)
   Pipette, 10 ml, with bulb
   Spatula
   Stirring bar
   Test tubes and caps (6)
   Test tube rack
   Wax pencil
   Weigh paper

Procedure

First Session

Media Preparation and Sterilization

1. Following the directions on the bottle, prepare 10 ml of tryptic soy broth in a small flask or beaker. Heat the mixture until the powder dissolves, and then transfer 5 ml to each of two tubes with a 10 ml pipette. Cap the tubes loosely, and place them in a test tube rack.

2. Following the directions on the bottle, prepare 115 ml of tryptic soy agar in a 250 ml Erlenmeyer flask. Heat the mixture until the powder dissolves and the medium turns clear. *Note:* To keep the powder from burning, swirl the flask contents occasionally. Watch your flask closely, since the medium will boil over if heated too long.

3. After the medium has become clear, use a 10 ml pipette to transfer 5 ml to each of two tubes and 10 ml to each of two other tubes. Cap loosely, and add these four tubes to the two already in the test tube rack. Leave the remaining 85 ml of tryptic soy agar in the flask, and cover with foil. This volume will be used to pour plates after sterilization.

4. Label your test tube rack and flask, and place them in the autoclave for sterilization.

5. After sterilization, lean the two tubes with 5 ml of tryptic soy agar against a notebook or similar object. When the medium cools, these tubes will form agar slants. Let the remaining tubes cool upright in the rack. These tubes will form agar deeps.

6. Let the flask contents cool sufficiently to allow handling without discomfort. When this has occurred, pour three plates using the technique depicted in figure 17.4. **Do not pour hot agar into a petri dish**. This will cause excess condensation on the lid of the petri dish and on the agar surface. Excess moisture may allow the culture to spread across the entire surface of the agar, instead of forming discrete colonies.

7. When the agar has gelled in tubes and plates, the media are ready to inoculate.

(a) Remove foil top and flame opening of flask.

(b) Pour sterile medium into petri dish to fill bottom. Allow the agar to cool and gel before moving dishes.

Figure 17.4 Preparation of agar plates.

Media Inoculation and Incubation

1. After the different forms of media (broth, slants, deeps, and plates) have been prepared, sterilized, and cooled, they are ready for inoculation. Begin by inoculating the two tubes of tryptic soy broth with a culture of *Escherichia coli*. Use an inoculating loop, and follow the procedure in figure 17.5.

(a) Label the tube to be inoculated with the microorganism used, the date, and your name or initials.

Culture

(b) Take the broth culture in one hand.

(c) Take the inoculating loop with your other hand, and flame the entire wire portion to redness.

(d) Remove the plug or cap from the tube by grasping it between the fingers of the hand holding the inoculating loop.

(e) Flame the mouth (lip) of the broth culture.

Culture

(f) Insert the sterile loop into the broth culture, and obtain a loopful of culture. Withdraw the loop, flame the mouth, and replace the plug or cap. Set the broth culture in a test tube rack. Pick up the tube to be inoculated, remove the plug or cap, and flame the mouth.

(g) Introduce the loopful of culture by immersing the loop into the sterile broth. Stir lightly.

(h) Withdraw the loop, flame the lip, and replace the plug or cap. Set the inoculated tube in the test tube rack.

(i) Flame the inoculating loop again and put it down. Incubate the inoculated tube as directed.

Figure 17.5 Broth inoculation.

2. Inoculate the two tryptic soy agar slants with *Escherichia coli* using an inoculating loop. Follow the procedure outlined in figure 17.6.

3. Inoculate the two tryptic soy agar deeps with *Escherichia coli* using an inoculating needle. The inoculation of a deep is depicted in figure 17.7.

(a) Label the agar slant to be inoculated with the microorganism to be used, the date, and your name or initials.

Culture

(b) Take the broth culture in one hand.

(c) Take the inoculating loop with your other hand, and flame the entire wire portion to redness.

(d) Remove the plug or cap from the tube by grasping it between the fingers of the hand holding the inoculating loop.

(e) Flame the mouth of the broth culture.

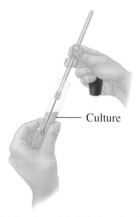

Culture

(f) Obtain a loopful of the broth culture. Withdraw the loop, flame the mouth, and replace the plug or cap. Set the tube down, and pick up the agar slant to be inoculated. Remove the plug or cap, and flame the mouth.

Agar slant

(g) Place the loop on the agar slant's surface at its bottom. Move the loop from side to side as you pull it upward out of the tube.

(h) Withdraw the loop, flame the lip of the tube, and replace the plug or cap. Place the tube in a rack, and incubate as directed.

(i) Flame the inoculating loop again, and put it down.

Figure 17.6 Inoculation of an agar slant.

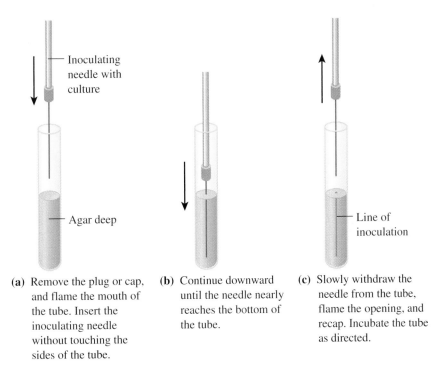

Inoculating needle with culture

Agar deep

Line of inoculation

(a) Remove the plug or cap, and flame the mouth of the tube. Insert the inoculating needle without touching the sides of the tube.

(b) Continue downward until the needle nearly reaches the bottom of the tube.

(c) Slowly withdraw the needle from the tube, flame the opening, and recap. Incubate the tube as directed.

Figure 17.7 Stab technique for agar deep cultures. The inoculating needle is sterilized and dipped into a broth culture before this step.

4. Inoculate the three plates of tryptic soy agar with *Escherichia coli* using the method depicted in figure 17.8. Spread the culture over several quadrants of the plate using the streak-plate method depicted in figure 17.9.

5. After inoculation, incubate plates and tubes for 48–72 hours in a laboratory incubator (figure 17.10) set at 35°C.

Second Session: Media Examination

1. After incubation, examine cultures for growth. *Note:* The assessment of growth in tubes and plates can be aided by comparing inoculated media with uninoculated media. Begin your assessment with broth tubes. Broth tubes should appear cloudy, or turbid, when compared to the clear broth in uninoculated tubes.

2. Growth on slants should be evident extending away from the line of inoculation.

3. In agar deeps, growth should occur along the needle line of inoculation from top to bottom.

4. Growth on agar plates should appear as distinct colonies by the second or third quadrant, each colony having the same appearance. Good separation of colonies is essential for two reasons: (1) to confirm the presence of only one species of bacteria (a pure culture); and (2) to determine the specific characteristics of isolated colonies. If colonies are not well separated, you might consider inoculating another series of plates in an attempt to improve your technique.

5. After examining all plates and tubes for *Escherichia coli* growth, go back through and inspect them for signs of contamination. Look for any type of growth that appears different from that of the culture you inoculated. The absence of contamination indicates good aseptic technique. Contamination in one or more tubes or plates may indicate the need to review the basic aseptic techniques in table 17.2.

(a) Flame the opening of the broth tube.

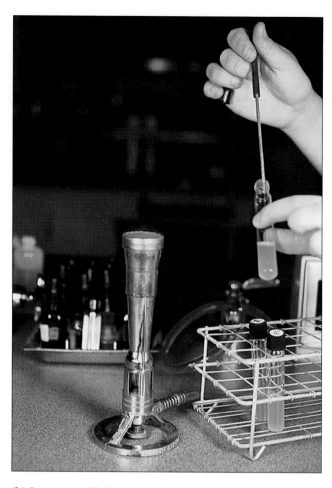

(b) Insert a sterile loop into the broth culture.

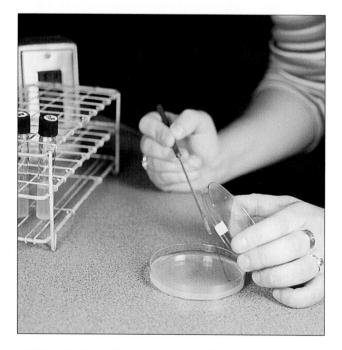

(c) Transfer the culture to a section of an agar plate by rubbing the loop back and forth across the surface.

Figure 17.8 Inoculation of an agar plate.

(a) Orient your plate as depicted here.

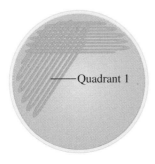

(b) Lift the lid, and use a sterile loop to make lines, or streaks, across the agar as shown in quadrant 1. Close the lid.

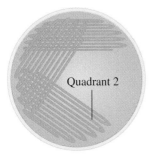

(c) Flame the loop, lift the lid, and make streaks as shown in quadrant 2. Close the lid.

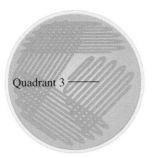

(d) Flame the loop, lift the lid, and make streaks as shown in quadrant 3. Close the lid. Incubate the plate as directed.

Figure 17.9 The streak-plate method.

Figure 17.10 A laboratory incubator. A temperature of 35°C will encourage rapid bacterial growth.

LABORATORY REPORT

NAME _____ DATE _____

LAB SECTION _____

Preparation and Inoculation of Growth Media

1. Draw the growth results from your tubes and plates.

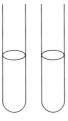 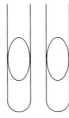 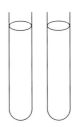

Broth tubes   Agar slants   Agar deeps

Growth (+ or −) _____ _____   _____ _____   _____ _____

Contaminants? _____ _____   _____ _____   _____ _____

Description
   of growth _____ _____   _____ _____   _____ _____

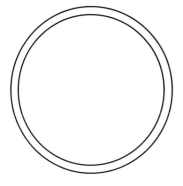 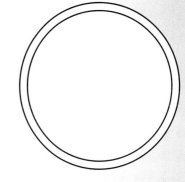

Agar plates

Growth (+ or −) _____   _____   _____

Contaminants? _____   _____   _____

Description
   of growth _____   _____   _____

Good separation
   of colonies? _____   _____   _____

2. Answer the following questions in the space provided.

   a. Why is it essential that media be sterile prior to use?

   b. Why must agar be cooled prior to pouring plates?

   c. Why are the inoculating loop and needle flamed before and after use?

   d. What is a contaminant? How does it gain entry into your culture? How do you keep a contaminant from entering your lab culture?

   e. How does one determine if growth has occurred in broth?

   f. Why is streak-plating such an essential procedure in the isolation and characterization of bacteria?

3. Answer the following questions based on these photographs:

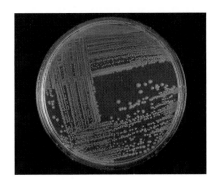

a. Name of technique _____

   Good separation of colonies?

   (yes or no) _____

   Pure culture? (yes or no) _____

   How do you know? _____

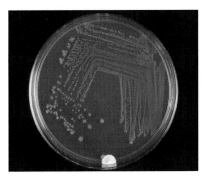

b. Name of technique _____

   Culture contaminant? (yes or no) _____

   How can you tell? _____

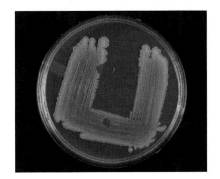

c. Name of technique _____

   Good separation of colonies?

   (yes or no) _____

   Pure culture? (yes or no) _____

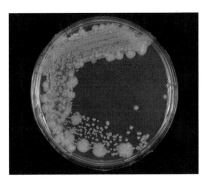

d. Name of technique _____

   Good separation of

   colonies? (yes or no) _____

   Pure culture? (yes or no) _____

   How do you know? _____

Culture Characterization of Bacteria

Background

When a single bacterial culture is grown using different forms of media (broth, slants, deeps, and plates), it displays a collective pattern of growth that is unique to its species. This unique pattern of growth is referred to as its **culture characteristics**. An organism's culture characteristics can help distinguish it from other organisms, since each bacterial species typically has a unique pattern of growth. Although useful, culture characteristics alone cannot be relied on to identify the many species of bacteria. They must be combined with staining reactions and biochemical characteristics.

In this exercise, you will use different forms of media to determine the culture characteristics of five known bacteria. You will also be given one of these five bacteria as an unknown to identify.

Materials

Cultures (24–48-hour broth)
Bacillus cereus
Micrococcus luteus
Proteus vulgaris
Pseudomonas aeruginosa
Staphylococcus epidermidis

Media
6 tubes tryptic soy broth
6 slants tryptic soy agar
6 deeps tryptic soy agar
6 plates tryptic soy agar
6 tubes motility test agar

Equipment
Incubator (set at 35°C)

Miscellaneous supplies
Bunsen burner and striker
Inoculating loop and needle
Test tube rack
Wax pencil

Procedure

First Session: Media Inoculation and Incubation

1. Inoculate 6 tryptic soy broth tubes: 5 tubes with the known cultures (*Bacillus cereus, Micrococcus luteus, Proteus vulgaris, Pseudomonas aeruginosa*, and *Staphylococcus epidermidis*) and 1 tube with the unknown culture (one of the previous five cultures, but designated by number only).

2. Inoculate 6 tryptic soy agar slants: 5 slants with the known cultures and 1 slant with the unknown culture.

3. Inoculate 6 tryptic soy agar deeps: 5 deeps with the known cultures and 1 deep with the unknown culture. Use an inoculating needle and a straight line of inoculation almost to the bottom.

4. Inoculate 6 motility test agar tubes: 5 tubes with the known cultures and 1 tube with the unknown culture. Use an inoculating needle and a straight line of inoculation two-thirds of the way down. **Motility test agar** is used to determine whether or not a culture is motile. The composition of this medium is listed in table 18.1.

Table 18.1 Components of Motility Test Agar

Tryptose	10 g
Sodium chloride	5 g
Agar	5 g
Distilled water	1,000 ml
Final pH =	7.2

Source: *The Difco Manual*. Eleventh Edition. Difco Laboratories.

5. Inoculate 6 tryptic soy agar plates: 5 plates with the known cultures and 1 plate with the unknown culture. Use an inoculating loop and the streak-plate method.

6. Incubate all inoculated tubes and plates in a 35°C incubator for 48–72 hours.

Second Session: Media Examination

1. After incubation, examine all plates and tubes for growth. *Note:* To aid in the interpretation of growth, use an uninoculated plate or tube for comparison. Begin your examination with the six broth tubes. Of the growth patterns in broth depicted in figure 18.1, determine which pattern is displayed by each culture. Record your determination in the table of the laboratory report.

2. Continue your examination of growth by inspecting slants, deeps, motility test agar, and plates. Again, consult the growth patterns in figure 18.1 to determine which pattern is displayed in the appropriate medium by each culture. Record your results in the table of the laboratory report.

3. After inspecting your cultures and completing the laboratory report table, determine which of the five cultures you have as your unknown.

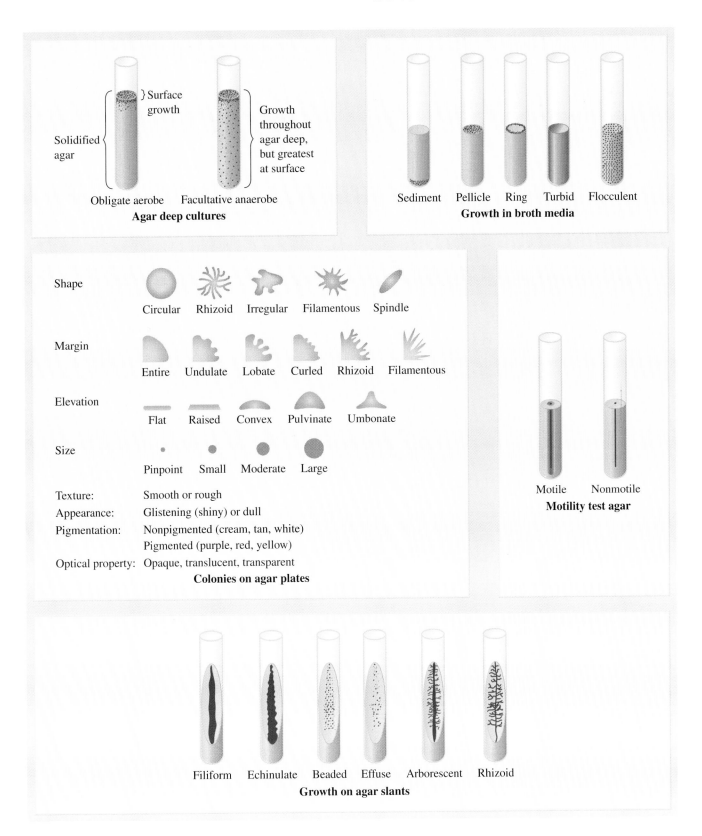

Figure 18.1 Cultural characteristics of bacteria.

LABORATORY REPORT

18

NAME _____ DATE _____

LAB SECTION _____

Culture Characterization of Bacteria

1. a. Fill in the following table from your observations of culture characteristics.

Organism	Colony morphology	Growth on slants	Growth in deeps	Growth in broth	Motility test agar
Bacillus cereus					
Micrococcus luteus					
Proteus vulgaris					
Pseudomonas aeruginosa					
Staphylococcus epidermidis					
Unknown no._____					

b. Based on the results you recorded in the table, identify your unknown: _____

_____

c. Which culture characteristic(s) were most useful to you in identifying your unknown?

d. Which organism has the following culture characteristics?

(1) water-soluble green pigment, forms pellicle in broth, and is motile: _____

(2) small-to-medium white colony, facultatively anaerobic, and nonmotile: _____

2. Does each culture appear to have its own unique culture characteristics? If so, explain how this could be useful in identification.

3. Define these terms:

   a. colony

   b. pigmentation

   c. facultatively anaerobic

   d. pellicle

4. Answer the following questions based on these photographs:

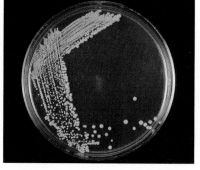
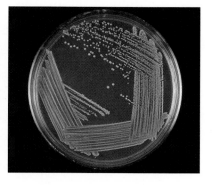

a. Name this growth
   pattern in broth.

   _____

   In this exercise,

   displayed by _____

b. Is there pigmentation? _____
   In this exercise, displayed by

   _____

c. A streak-plate of which
   organism in this exercise?

   _____

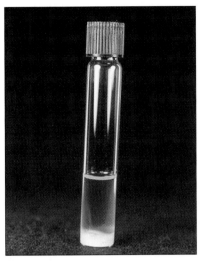

d. A culture characteristic of which organism in this

exercise? _____

e. Name this growth pattern

in broth. _____

In this exercise, displayed

by _____

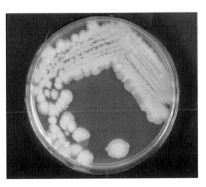

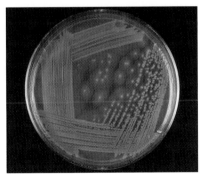

f. Is there pigmentation? _____ In this exercise, this colony morphology displayed by

_____

g. A streak-plate of which organism in this exercise?

_____

Bacterial Identification

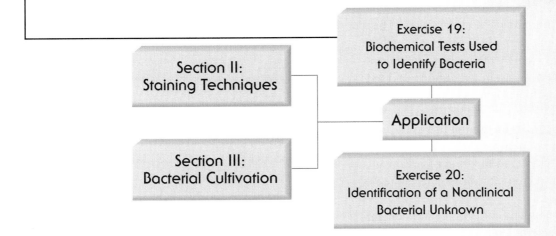

Section II:
Staining Techniques

Section III:
Bacterial Cultivation

Exercise 19:
Biochemical Tests Used
to Identify Bacteria

Application

Exercise 20:
Identification of a Nonclinical
Bacterial Unknown

19

Biochemical Tests Used to Identify Bacteria

Background

Although culture and staining characterization of bacteria provide a substantial amount of information, these techniques are not sufficient by themselves for the identification of bacteria. The results of staining and culturing must be combined with the results of **biochemical tests** to definitively identify bacteria. Biochemical tests evaluate the metabolic properties of a bacterial isolate. After a number of biochemical tests have been performed, the combination of test results forms a biochemical pattern for an isolate, which is unique for each species.

In this exercise, you will perform eight biochemical tests on known bacterial cultures. You will use two cultures for each test, a culture known to have a positive result and a culture known to have a negative result. This will familiarize you with either test result, allowing you to correctly interpret biochemical test results for the nonclinical unknown you will identify in Exercise 20.

Materials

Cultures (24–48-hour agar or broth)
Alcaligenes faecalis
Enterobacter aerogenes
Enterococcus faecalis
Escherichia coli
Proteus vulgaris
Pseudomonas aeruginosa
Staphylococcus epidermidis

Media
1 plate tryptic soy agar
2 slants tryptic soy agar
6 tubes oxidation-fermentation (O-F) glucose
   medium
2 tubes nitrate broth (with durham tube)
2 tubes methyl red-Voges Proskauer (MR-VP)
   medium
4 tubes sulfide indole motility (SIM) medium
2 tubes lactose broth (with durham tube)

Equipment
Incubator (set at 35°C)

Reagents
Hydrogen peroxide (3%)
Kovac's reagent
Methyl red (pH indicator)
Oxidase reagent

Miscellaneous supplies
Bunsen burner and striker
Inoculating loop and needle
Mineral oil (sterile)
Pasteur pipette with bulb
Test tube rack
Wax pencil

Procedure

First Session: Inoculation and Incubation

1. *Catalase test:* Inoculate 2 tryptic soy agar slants, one with *Enterococcus faecalis,* and the other with *Staphylococcus epidermidis.* Use an inoculating loop to make a back-and-forth streak across the slant surface.

2. *Denitrification test:* Using an inoculating loop, inoculate 2 nitrate broth tubes, one with *Alcaligenes faecalis,* and the other with *Pseudomonas aeruginosa.*

3. *Hydrogen sulfide (H_2S) production:* Inoculate 2 SIM tubes, one with *Escherichia coli,* and the other with *Proteus vulgaris.* Use an inoculating needle, and stab the agar with a single in-and-out motion.

4. *Indole production:* Inoculate 2 SIM tubes, one with *Enterobacter aerogenes,* and the other with *Escherichia coli.* Use an inoculating needle, and stab the agar with a single in-and-out motion.

5. *Lactose utilization:* Using an inoculating loop, inoculate 2 lactose broth tubes, one with *Escherichia coli*, and the other with *Proteus vulgaris.*

6. *Methyl red test:* Using an inoculating loop, inoculate 2 MR-VP tubes, one with *Enterobacter aerogenes,* and the other with *Escherichia coli.*

7. *Oxidase test:* With a wax pencil, draw a line down the center of a tryptic soy agar plate. Using an inoculating loop, inoculate one half of the plate with *Escherichia coli* and the other half with *Pseudomonas aeruginosa.* Use a back-and-forth streak across the surface of the agar.

8. *Oxidation-fermentation (O-F) glucose test:* Using an inoculating needle, inoculate 2 tubes of O-F glucose with *Alcaligenes faecalis,* 2 tubes with *Escherichia coli,* and 2 tubes with *Pseudomonas aeruginosa.* After inoculation, cover one tube in each pair with a 2 cm layer of sterile mineral oil. *Note:* Mineral oil can be poured directly into the tube without using a pipette.

9. Incubate all tubes and plates at 35°C for 24–48 hours, except MR-VP. MR-VP tubes require a minimum of 72 hours of incubation.

Second Session: Reading Test Results

1. *Catalase test:* Use a Pasteur pipette to place a few drops of 3% hydrogen peroxide onto each slant culture. Watch for immediate signs of bubbling, which represent a positive test; the absence of bubbles is a negative test (figure 19.1). A slide test can be done by mixing a small amount of culture into a drop of water on a glass slide. The hydrogen peroxide is then added to the drop.

   Expected results: Staphylococcus epidermidis is catalase-positive, while *Enterococcus faecalis* is catalase-negative.

2. *Denitrification test:* Note the small inverted tube in the bottom of the medium. This tube, called a durham tube, is designed to collect gas. Read this test by looking for gas bubbles in the durham

Positive test:

$$\text{added } H_2O_2 \xrightarrow{\text{catalase}} H_2O + O_2$$
(bubbles)

Example:
Staphylococcus epidermidis

Negative test:

$$\text{added } H_2O_2 \xrightarrow{\text{no catalase}} H_2O_2$$
(no bubbles)

Example:
Enterococcus faecalis

Test results in tubes

Positive test result on a slide

Figure 19.1 Catalase test: reactions and results of positive and negative tests.

tube (nothing needs to be added). Nitrate broth contains potassium nitrate (table 19.1). Denitrification by bacteria converts the nitrate to nitrogen gas. Gas bubbles in the durham tube, therefore, represent a positive test (figure 19.2). The absence of bubbles represents a negative test.

Table 19.1 Composition of Biochemical Test Media

Nitrate broth

Peptone	5 g
Beef extract	3 g
Potassium nitrate	1 g
Distilled water	1,000 ml
Final pH = 7.0	

SIM medium

Peptone	30 g
Beef extract	3 g
Peptonized iron	0.2 g
Sodium thiosulfate	0.02 g
Agar	3 g
Distilled water	1,000 ml
Final pH = 7.3	

Lactose broth

Beef extract	1 g
Proteose peptone	10 g
Sodium chloride	5 g
Lactose	5 g
Phenol red	0.018 g
Distilled water	1,000 ml
Final pH = 7.4	

MR-VP medium

Peptone	7 g
Dextrose	5 g
Dipotassium phosphate	5 g
Distilled water	1,000 ml
Final pH = 6.9	

Oxidation-fermentation (O-F) glucose medium

Glucose	10 g
Tryptone	2 g
Sodium chloride	5 g
Dipotassium phosphate	0.3 g
Bromthymol blue	0.08 g
Agar	2 g
Distilled water	1,000 ml
Final pH = 6.8	

Source: *The Difco Manual.* Eleventh Edition. Difco Laboratories.

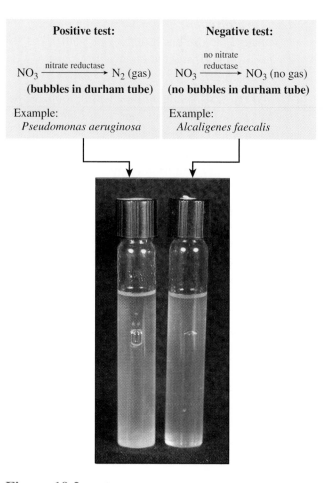

Positive test:	Negative test:
$NO_3 \xrightarrow{\text{nitrate reductase}} N_2$ (gas) **(bubbles in durham tube)**	$NO_3 \xrightarrow[\text{reductase}]{\text{no nitrate}} NO_3$ (no gas) **(no bubbles in durham tube)**
Example: *Pseudomonas aeruginosa*	Example: *Alcaligenes faecalis*

Figure 19.2 Denitrification: positive and negative test results.

Expected results: Pseudomonas aeruginosa is positive, while *Alcaligenes faecalis* is negative.

3. ***Hydrogen sulfide (H₂S) production:*** Examine each SIM tube for the presence of a black color (nothing needs to be added). A black color indicates the production of H_2S, which combines with the peptonized iron in the SIM medium (table 19.1). The result is FeS, which causes a blackening of the medium and represents a positive test (figure 19.3). The absence of a black color is a negative test.

Expected results: Proteus vulgaris is positive, while *Escherichia coli* is negative.

4. ***Indole production:*** Use a dropper to place 5 drops of Kovac's reagent onto the top of the SIM agar in each tube. If the amino acid tryptophan has been broken down by the enzyme tryptophanase to form indole, the Kovac's reagent will combine with the indole to form a red color. A red color in the Kovac's reagent at the top of the agar represents a positive test (figure 19.4).

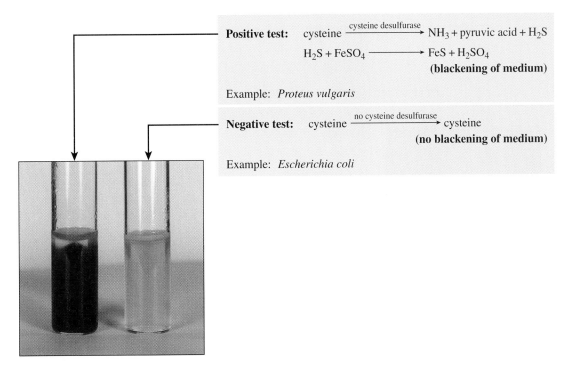

Positive test: cysteine $\xrightarrow{\text{cysteine desulfurase}}$ NH$_3$ + pyruvic acid + H$_2$S

$\text{H}_2\text{S} + \text{FeSO}_4 \longrightarrow \text{FeS} + \text{H}_2\text{SO}_4$
(blackening of medium)

Example: *Proteus vulgaris*

Negative test: cysteine $\xrightarrow{\text{no cysteine desulfurase}}$ cysteine
(no blackening of medium)

Example: *Escherichia coli*

Figure 19.3 Hydrogen sulfide (H$_2$S) production: positive and negative test results.

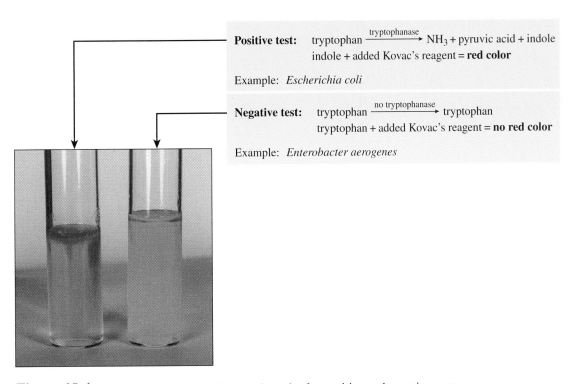

Positive test: tryptophan $\xrightarrow{\text{tryptophanase}}$ NH$_3$ + pyruvic acid + indole
indole + added Kovac's reagent = **red color**

Example: *Escherichia coli*

Negative test: tryptophan $\xrightarrow{\text{no tryptophanase}}$ tryptophan
tryptophan + added Kovac's reagent = **no red color**

Example: *Enterobacter aerogenes*

Figure 19.4 Indole production: reactions and results for positive and negative tests.

No color change in the Kovac's reagent is a negative test.

Expected results: *Escherichia coli* is indole-positive, while *Enterobacter aerogenes* is indole-negative.

5. ***Lactose utilization:*** When examining these tubes, look for a color change in the broth and gas in the durham tube (nothing needs to be added). Lactose broth contains the sugar lactose and the pH indicator phenol red (table 19.1). When lactose is

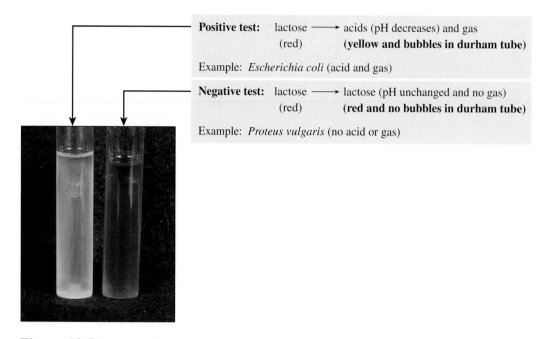

Figure 19.5 Lactose utilization: possible reactions and results.

utilized, acids or acids and gas are produced. The acid causes the pH to decrease, turning the phenol red from red to yellow. The gas collects in the durham tube. Therefore, a yellow color or a yellow color and bubbles in the durham tube represent a positive test (figure 19.5). No color change and no bubbles in the durham tube represent a negative test.

Expected results: *Escherichia coli* is positive, while *Proteus vulgaris* is negative.

6. ***Methyl red test:*** Using a Pasteur pipette, add 10 drops of methyl red pH indicator to each tube. Swirl the tube gently to mix the drops into the broth. Examine each tube for color change. Bacteria that produce many acids from the breakdown of dextrose (glucose) in the MR-VP medium (table 19.1) cause the pH to drop to 4.2. At this pH, methyl red is red. A red color represents a positive test (figure 19.6). Bacteria that produce fewer acids from the breakdown of glucose drop the pH to only 6.0. At 6.0, methyl red is yellow. A yellow color represents a negative test.

Expected results: *Escherichia coli* is methyl-red-positive, while *Enterobacter aerogenes* is methyl-red-negative.

7. ***Oxidase test:*** Drop 1–2 drops of oxidase reagent onto colonies of both cultures. Watch for a gradual color change from pink, to light purple, and then to dark purple within 10–30 seconds. Such a color change indicates the presence of the respiratory enzyme cytochrome *c* oxidase and represents a positive test (figure 19.7). No color change in this period is a negative test.

Expected results: *Pseudomonas aeruginosa* is oxidase-positive, while *Escherichia coli* is oxidase-negative.

8. ***Oxidation-fermentation (O-F) glucose test:*** In these tubes, you will look for color changes in the medium (nothing needs to be added). O-F glucose medium contains the sugar glucose and the pH indicator bromthymol blue (table 19.1). This pH indicator is green at the initial pH of 6.8, but turns to yellow at a pH of 6.0. If glucose is utilized, acids are produced and the pH drops,

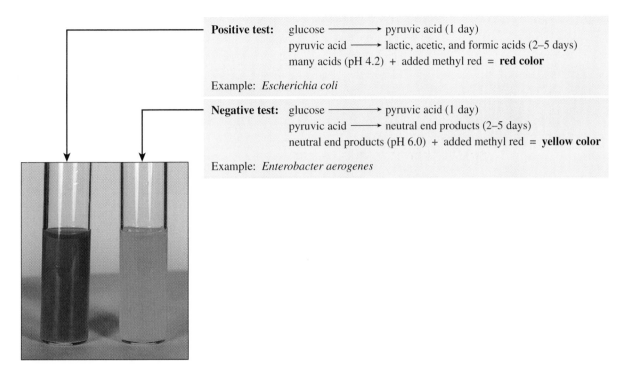

Positive test: glucose ⟶ pyruvic acid (1 day)
                      pyruvic acid ⟶ lactic, acetic, and formic acids (2–5 days)
                      many acids (pH 4.2) + added methyl red = **red color**

Example: *Escherichia coli*

Negative test: glucose ⟶ pyruvic acid (1 day)
                      pyruvic acid ⟶ neutral end products (2–5 days)
                      neutral end products (pH 6.0) + added methyl red = **yellow color**

Example: *Enterobacter aerogenes*

Figure 19.6 Methyl red test: reactions and results for a positive and negative test.

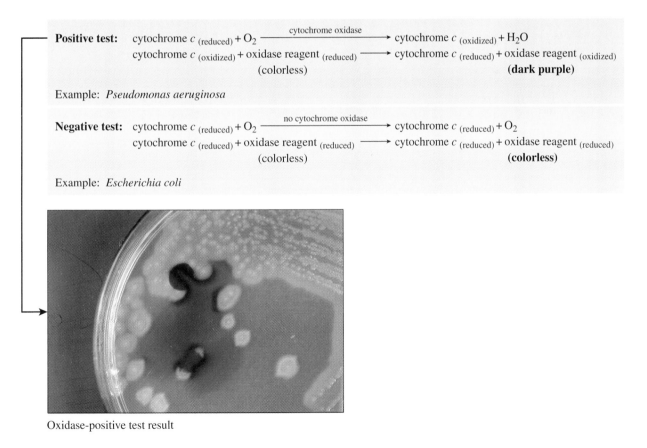

Positive test: cytochrome $c_{(reduced)}$ + O_2 $\xrightarrow{\text{cytochrome oxidase}}$ cytochrome $c_{(oxidized)}$ + H_2O
                      cytochrome $c_{(oxidized)}$ + oxidase reagent $_{(reduced)}$ ⟶ cytochrome $c_{(reduced)}$ + oxidase reagent $_{(oxidized)}$
                      (colorless)   **(dark purple)**

Example: *Pseudomonas aeruginosa*

Negative test: cytochrome $c_{(reduced)}$ + O_2 $\xrightarrow{\text{no cytochrome oxidase}}$ cytochrome $c_{(reduced)}$ + O_2
                      cytochrome $c_{(reduced)}$ + oxidase reagent $_{(reduced)}$ ⟶ cytochrome $c_{(reduced)}$ + oxidase reagent $_{(reduced)}$
                      (colorless)   **(colorless)**

Example: *Escherichia coli*

Oxidase-positive test result

Figure 19.7 Oxidase test: reactions and results of positive and negative tests.

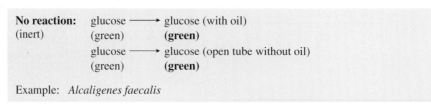

No reaction: glucose ⟶ glucose (with oil)
(inert) (green) **(green)**
glucose ⟶ glucose (open tube without oil)
(green) **(green)**

Example: *Alcaligenes faecalis*

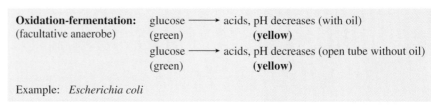

Oxidation-fermentation: glucose ⟶ acids, pH decreases (with oil)
(facultative anaerobe) (green) **(yellow)**
glucose ⟶ acids, pH decreases (open tube without oil)
(green) **(yellow)**

Example: *Escherichia coli*

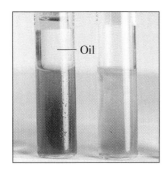
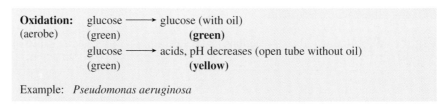

Oxidation: glucose ⟶ glucose (with oil)
(aerobe) (green) **(green)**
glucose ⟶ acids, pH decreases (open tube without oil)
(green) **(yellow)**

Example: *Pseudomonas aeruginosa*

Figure 19.8 Oxidation-fermentation (O-F) glucose test: possible reactions and results.

causing the bromthymol blue to turn from green to yellow. If both tubes (with and without oil) turn yellow, the test organism is considered a facultative anaerobe, able to use glucose in the presence or absence of oxygen (figure 19.8). If only the tube without oil turns yellow, the test organism is considered an aerobe, able to use glucose only when oxygen is present. No change in either tube indicates that the test organism is unable to utilize glucose.

Expected results: *Escherichia coli* is a facultative anaerobe, *Pseudomonas aeruginosa* is an aerobe, and *Alcaligenes faecalis* is nonreactive (inert) on glucose.

LABORATORY REPORT

NAME _____ DATE _____

LAB SECTION _____

Biochemical Tests Used to Identify Bacteria

Record your results for the biochemical tests.

Biochemical test	Reagent added	Observations	Interpretation
1. Catalase test: *Enterococcus faecalis*			
Staphylococcus epidermidis			
2. Denitrification test: *Alcaligenes faecalis*			
Pseudomonas aeruginosa			
3. H$_2$S production: *Escherichia coli*			
Proteus vulgaris			
4. Indole production: *Enterobacter aerogenes*			
Escherichia coli			
5. Lactose utilization: *Escherichia coli*			
Proteus vulgaris			
6. Methyl red test: *Enterobacter aerogenes*			
Escherichia coli			
7. Oxidase test: *Escherichia coli*			
Pseudomonas aeruginosa			
8. O-F glucose test: *Alcaligenes faecalis*			
Pseudomonas aeruginosa			
Escherichia coli			

Application: Identification of a Nonclinical Bacterial Unknown

Background

Previous exercises have covered basic aspects of microbiology, including microscopic observation, staining, cultivation, and biochemical testing. In this exercise, you will apply what you have learned about these techniques to the identification of a nonclinical bacterial unknown.

Materials

Cultures (24–48-hour broth)
  Alcaligenes faecalis
  Bacillus cereus
  Enterobacter aerogenes
  Enterococcus faecalis
  Escherichia coli
  Micrococcus luteus
  Mycobacterium phlei
  Neisseria sicca
  Proteus vulgaris
  Pseudomonas aeruginosa
  Serratia marcescens
  Staphylococcus epidermidis

 All agents in red are BSL2 bacteria.

Stains
  Gram stain
    Crystal violet
    Gram's iodine
    Ethanol (95%)
    Safranin

  Acid-fast stain
    Carbolfuchsin
    Acid-alcohol
    Methylene blue

  Spore stain
    Malachite green
    Safranin

Media
  Tryptic soy agar plates
  Tryptic soy agar slants
  O-F glucose tubes
  Nitrate broth tubes (with durham tube)
  MR-VP tubes
  SIM tubes
  Lactose broth tubes (with durham tube)

Equipment
  Hot plate (optional)
  Incubator (set at 35°C)
  Light microscope

Reagents
  Hydrogen peroxide (3%)
  Kovac's reagent
  Methyl red (pH indicator)
  Oxidase reagent

Miscellaneous supplies
  Bibulous paper
  Bunsen burner and striker
  Clothespin
  Disposable gloves (optional)
  Egg albumin solution
  Glass slides
  Immersion oil
  Inoculating loop and needle
  Lens paper
  Mineral oil (sterile)
  Pasteur pipette with bulb
  Staining tray
  Test tube rack
  Wash bottle with tap water
  Wax pencil

Procedure

1. You will select an unknown bacterial culture, or one will be assigned to you. In either case, be sure to record in the laboratory report the number assigned to your unknown.

2. Do a streak-plate of your unknown (see Exercise 17). After incubation, examine your streak-plate to make sure that you have a pure culture free of contamination.

3. Examine the streak-plate to determine the colony characteristics of your unknown culture (see Exercise 18). Also examine the broth tube your unknown was cultured in to determine its characteristics (see Exercise 18).

4. Do a Gram stain of an 18–24-hour culture to determine cell morphology and Gram reaction (see Exercise 11). Cell shape and arrangement can be verified with either a negative stain (see Exercise 9) or a simple stain (see Exercise 10).

5. Examine the identification scheme in figure 20.1 to determine the test to be done next. If a spore stain is required, consult Exercise 13; if an acid-fast stain is necessary, see Exercise 12; if biochemical tests are needed, you can find them in Exercise 19. *Note:* Inoculate a new plate each week to keep your culture viable.

6. Continue with your tests until your unknown has been identified. Be sure to record the results of all tests and the identity of your unknown in the laboratory report.

7. Your laboratory instructor may wish to see all results when you are finished. Therefore, be sure to keep all slides, plates, and tubes until examined by your laboratory instructor.

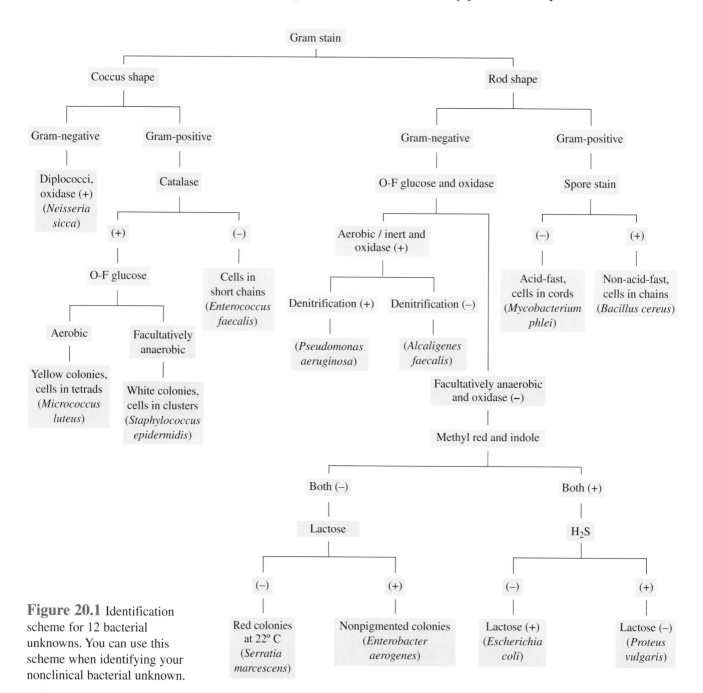

Figure 20.1 Identification scheme for 12 bacterial unknowns. You can use this scheme when identifying your nonclinical bacterial unknown.

LABORATORY REPORT

NAME _____ DATE _____

LAB SECTION _____

Application: Identification of a Nonclinical Bacterial Unknown

Unknown no. _____

1. Follow the identification scheme in figure 20.1 to identify your nonclinical bacterial unknown. Be sure to perform only the tests required to identify your unknown.

2. Record your results for the required tests.

Procedure	Observations	Results
Culture characteristics *Broth*		
Agar		
Staining characteristics *Cell shape*		
Cell arrangement		
Gram stain		
Acid-fast stain		
Spore stain		
Biochemical characteristics *Catalase test*		
Denitrification test		
H_2S production		
Indole production		
Lactose utilization		
Methyl red test		
Oxidase test		
O-F glucose test		

3. After following the scheme in figure 20.1 and recording the results for the required tests in the preceding table, I conclude that my unknown is _____.

Medical Microbiology

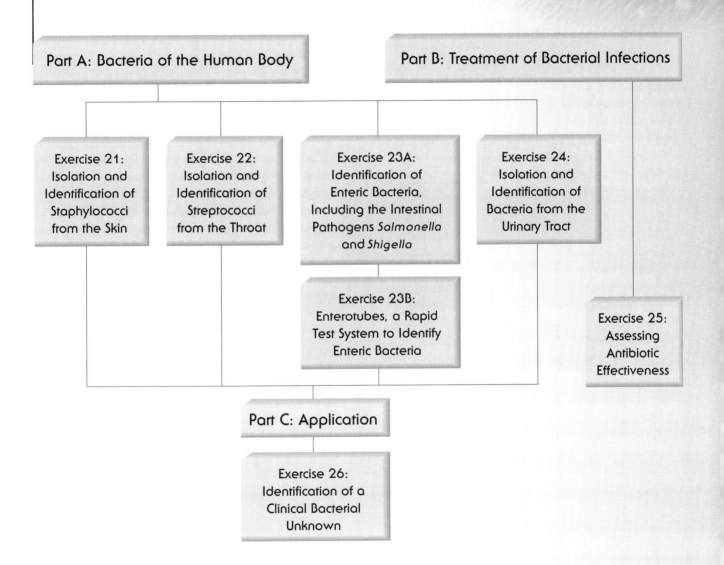

Part A: Bacteria of the Human Body

Part B: Treatment of Bacterial Infections

Exercise 21:
Isolation and
Identification of
Staphylococci
from the Skin

Exercise 22:
Isolation and
Identification of
Streptococci
from the Throat

Exercise 23A:
Identification of
Enteric Bacteria,
Including the Intestinal
Pathogens *Salmonella*
and *Shigella*

Exercise 24:
Isolation and
Identification of
Bacteria from the
Urinary Tract

Exercise 23B:
Enterotubes, a Rapid
Test System to Identify
Enteric Bacteria

Exercise 25:
Assessing
Antibiotic
Effectiveness

Part C: Application

Exercise 26:
Identification of a
Clinical Bacterial
Unknown

21

Isolation and Identification of Staphylococci from the Skin

Background

Normal Flora of Human Skin

Many regions of the body, including the skin, have a usual population of bacteria referred to as **residents** or **normal flora**. Particularly common on the skin are the Gram-positive cocci, including nonpathogenic *Staphylococcus epidermidis* and species of *Micrococcus*. In addition to these, Gram-positive pleomorphic rods, called diphtheroids, are also found.

Pathogens of Human Skin

Staphylococcus aureus is a normal resident of the nasal membranes, but can be transferred to the skin, where it is considered a transient. *S. aureus* is considered a **pathogen** because it causes skin infections such as boils, abscesses, carbuncles, impetigo, and scalded skin syndrome. In addition, exotoxin-producing strains of *S. aureus* cause food poisoning and toxic shock.

Identification of Skin Isolates

The Gram stain is used to determine the morphology and Gram reaction of skin isolates. The catalase test is used to differentiate the Gram-positive staphylococci and micrococci, which are catalase-positive (table 21.1),

from the Gram-positive streptococci, which are catalase-negative. Catalase-positive, Gram-positive cocci are differentiated using the oxidation-fermentation (O-F) glucose test (see Exercise 19), where the staphylococci are facultatively anaerobic (table 21.1). The presence of staphylococci is verified by growth on **mannitol salt agar** (MSA), since only these bacteria can tolerate the 7.5% salt content of the medium (table 21.2).

The identification of *Staphylococcus aureus* can be completed with the tests just described plus three additional tests. *S. aureus* ferments mannitol in MSA to produce acids that turn the medium from red to yellow. In addition, *S. aureus* produces **coagulase**, an enzyme that clots blood plasma, and **hemolysins**, enzymes that lyse red blood cells. The presence of the former enzyme is detected by a coagulase test, while the latter enzymes are detected with blood agar.

Detecting Penicillinase-Producing Staphylococci

The enzyme **penicillinase** opens the beta-lactam ring of the penicillin molecule, resulting in harmless penicilloic acid (figure 21.1). Therefore, bacteria able to produce penicillinase can break down penicillin and as a result are **penicillin-resistant**. With the extensive use of penicillin over 50 years, many bacteria are penicillin-

Table 21.1 Characteristics of Common Gram-positive Cocci from the Skin

Characteristic	*Micrococcus*	*Staphylococcus epidermidis*	*Staphylococcus aureus*
Pigment on agar	Bright yellow	White	Light to golden yellow
Catalase	(+)	(+)	(+)
O-F glucose	Aerobic	Facultatively anaerobic	Facultatively anaerobic
Mannitol fermentation	(−)	(−)	(+)
Coagulase	(−)	(−)	(+)
Hemolysis on blood agar	None	None	Beta-hemolysis

Table 21.2 The Composition of Mannitol Salt Agar (MSA)

Proteose peptone	10 g
Beef extract	1 g
D-mannitol	10 g
Sodium chloride	75 g
Phenol red	0.025 g
Agar	15 g
Distilled water	1,000 ml
Final pH	7.4

Source: *The Difco Manual.* Eleventh Edition. Difco Laboratories.

resistant, including the majority of staphylococci. Isolating staphylococci from the skin offers an excellent opportunity to demonstrate penicillin resistance. This can be easily done by using nitrocefin-impregnated disks. Nitrocefin has a beta-lactam ring similar to that of penicillin. When the ring of nitrocefin is opened by penicillinase, the molecule turns red. Therefore, the appearance of a red color on a nitrocefin dry slide after the addition of a staphylococcal skin isolate is indicative of the enzyme penicillinase.

Materials

Cultures (24–48-hour agar)
  Staphylococcus aureus
  Staphylococcus epidermidis

 All agents in red are BSL2 bacteria.

Media
  Blood agar plates (tryptic soy agar with 5% sheep blood)
  O-F glucose tubes
  Mannitol salt agar (MSA) plates
  Tryptic soy agar plates
  Tryptic soy broth tubes

Chemicals and reagents
  Blood plasma (rabbit; for coagulase test)
  Gram-stain reagents
  Hydrogen peroxide (for catalase test)
  Nitrocefin dry slides
  Rapid latex agglutination test kit (for coagulase test)

Equipment
  Incubator (35°C)
  Light microscope

Miscellaneous supplies
  Bibulous paper
  Biohazard bag (or similar container)
  Bottle with tap water
  Bunsen burner and striker
  Cotton-tipped swabs, sterile
  Disposable gloves
  Glass microscope slides
  Immersion oil
  Inoculating loop and needle
  Lens paper
  Mineral oil, sterile
  Pasteur pipette with bulb
  Pipette, 1 ml
  Test tube
  Test tube rack
  Wax pencil

Figure 21.1 Penicillinase produces resistance to penicillin by breaking a bond (arrow) in the beta-lactam ring of penicillin, resulting in penicilloic acid, a molecule with no effect on bacterial growth.

Procedure

First Session: Isolation of Bacteria from the Skin

1. Following the procedure shown in figure 21.2, dip a sterile cotton-tipped swab into a tube of tryptic soy broth. Blot the excess fluid against the side of the tube.

2. Select a portion of your arm, face, or leg. Rub the swab back and forth across a 5-square-centimeter area.

3. Use the swab to inoculate a tryptic soy agar (TSA) plate. Rub the swab back and forth over one area of the plate near the edge (figure 21.2c). *Note:* Dispose of the swab in a biohazard bag. With a sterile loop, cross over the swabbed area to spread the bacteria across the plate surface. Repeat this spreading process with the loop in two more quadrants as depicted in figure 21.2e, f.

4. Place the TSA plate into a 35°C incubator.

Second Session

Selection of Skin Isolates

1. After 24–48 hours, examine the inoculated plate. Based on colony morphology, determine the total number of different bacterial types on TSA. Record this number in the laboratory report.

2. Select two different bacteria from the plate that are most common—that is, the bacterial types with the greatest number of colonies. Number these #1 and #2, and record their colony morphology along with the colony morphology of the common skin isolates, *Staphylococcus epidermidis* and *S. aureus*.

Identification of Skin Isolates

The two common skin isolates will be identified using the tests specified in figure 21.3. You will also test *Staphylococcus epidermidis* and *Staphylococcus aureus* in conjunction with your unknowns.

1. Gram-stain your unknown skin isolates and the two known cultures. Record your results in the laboratory report.

2. If one or both of your unknown isolates are Gram-positive cocci, continue your identification by doing a catalase test as follows: Use a sterile loop to deposit some cells from the unknown colony into a drop of water on a glass slide. Add a drop of hydrogen peroxide. Watch for bubbles, indicative of a positive test. Do the same for your two known cultures.

(a) Moisten a sterile cotton-tipped swab in tryptic soy broth.

(b) Swab an area of your skin.

(c) Inoculate a plate by rubbing the swab back and forth near the edge.

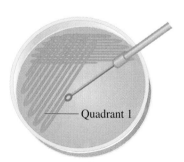

(d) Use a sterile loop to spread the inoculum into quadrant 1.

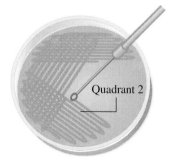

(e) Use a loop to spread into quadrant 2.

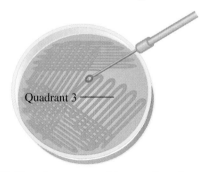

(f) Use a loop to spread into quadrant 3.

Figure 21.2 Isolation of bacteria from the skin.

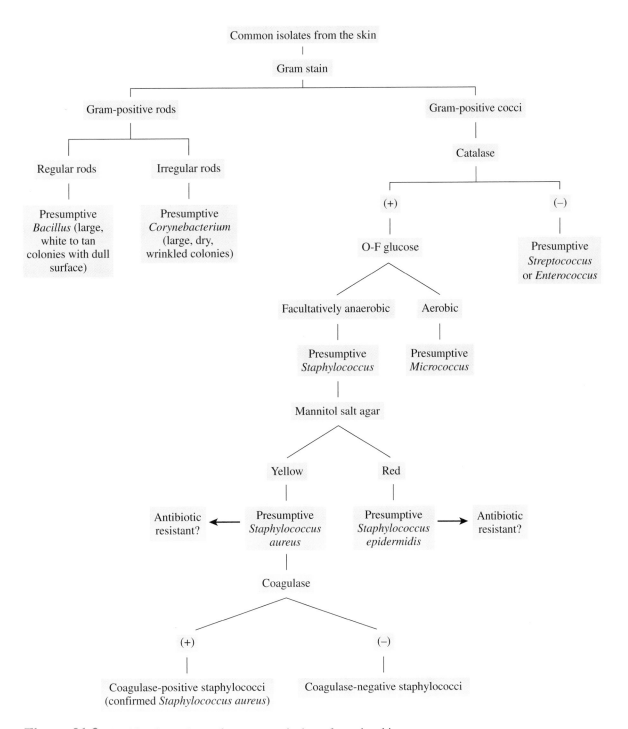

Figure 21.3 Identification scheme for common isolates from the skin.

3. If one or both of your unknown isolates is a Gram-positive coccus and catalase-positive, inoculate a mannitol salt agar (MSA) plate, a blood agar plate, and a pair of O-F glucose tubes (cover the medium in one tube with sterile mineral oil). Do the same for your two known cultures.

4. Incubate plates and tubes at 35°C.

Third Session

Identification of Staphylococci

1. After 24–48 hours, inspect each O-F tube for color change and each MSA plate for growth. The presence of growth on MSA and a color change from red to yellow in both O-F glucose tubes is indicative of staphylococci. If one or

more of your unknown isolates has these test results, you have confirmed the isolation of staphylococci from your skin. Both known cultures should yield these tests results.

2. To determine whether your isolate is the nonpathogenic *Staphylococcus epidermidis* or the pathogenic *Staphylococcus aureus*, examine the MSA plate for color change and the blood agar plate for hemolysis. *S. aureus* ferments mannitol to acids, yielding a color change from red to yellow. *S. aureus* also produces hemolysins, which lyse red blood cells, causing a clear zone of beta-hemolysis around the colonies. The nonpathogenic *S. epidermidis* produces no color change on MSA and no clear zone around colonies on blood agar.

3. To confirm the presence of *S. aureus* for those organisms that ferment mannitol and are beta-hemolytic, perform a coagulase test. This can be done using one of the following two methods. A positive result using either test provides confirmation of *S. aureus*.

Rabbit Plasma: Mix a loopful of the organism into 0.5 ml of rehydrated rabbit plasma in a test tube. Incubate the tube at 35°C for 4 hours. After incubation, examine the plasma for clotting by tilting the tube to the side. Plasma that has clotted will not run, indicating a positive test.

Latex agglutination: A rapid latex agglutination test kit can detect coagulase. Its use is outlined in figure 21.4.

Determination of Penicillin Resistance

1. To ascertain if your staphylococci are penicillin-resistant, take a loopful of culture from the MSA or blood agar plate, and rub it onto a moistened nitrocefin disk.

2. Examine the disk for color change. Penicillinase-producing staphylococci will yield a red color as the beta-lactam ring of nitrocefin is opened. Staphylococci that are penicillinase-negative produce no color change. Record your results in the laboratory report.

(a) Add one drop of latex reagent.

(b) Select an isolated colony, and pick it up with a stick.

(c) Mix the organism into the latex reagent.

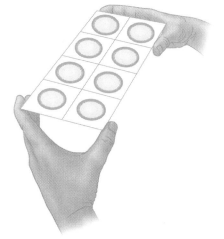

(d) Rotate the card gently after inoculation.

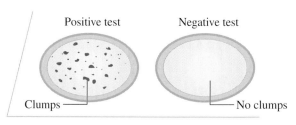

(e) Examine the card for black clumps (agglutination).

Figure 21.4 Use of the rapid latex agglutination test to detect coagulase.

LABORATORY REPORT

NAME _____ DATE _____

LAB SECTION _____

Isolation and Identification of Staphylococci from the Skin

1. Total number of bacterial types from the skin on TSA = _____

2. Identification of skin isolates

    a. **Identification of staphylococci**

Common skin isolates	Colony morphology	Cell morphology and Gram reaction	Catalase test	*Results indicative of staphylococci?
#1				
#2				
Known cultures: S. epidermidis				
S. aureus				

*If yes, continue with (b); if no, stop here.

    b. **Confirmation of staphylococci**

Common skin isolates	O-F glucose (facultatively anaerobic?)	Growth on MSA?	*Confirmed staphylococci?
#1			
#2			
Known cultures: S. epidermidis			
S. aureus			

*If yes, continue to (c); if no, stop here.

    c. **Differentiation of staphylococci**

Common skin isolates	Yellow on MSA?	Beta-hemolysis on blood agar?	Coagulase?	Indicative of S. epidermidis or S. aureus?
#1				
#2				
Known cultures: S. epidermidis				
S. aureus				

d. Did you isolate nonpathogenic *Staphylococcus epidermidis* from your skin? Is this a common result? Explain.

e. Did you isolate pathogenic *S. aureus* from your skin? Is this a common result? Explain.

3. Penicillin resistance

   a. Was your staphylococcal isolate penicillin-resistant (i.e., produced red color change on
      nitrocefin)? _____

   b. For your lab section, determine how many staphylococcal isolates were tested for penicillin resistance
      and how many of these were resistant. Calculate the percentage of penicillin-resistant isolates.

      Total tested = _____

      Total penicillin-resistant = _____

      Percent penicillin-resistant = _____

      Was this percentage expected? Explain.

4. Answer the following questions based on these photographs:

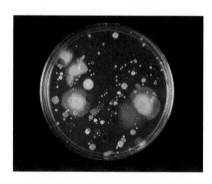 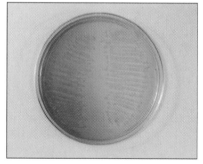 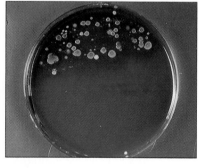

a. How many different bacterial types are on this skin plate?

_____

b. Are these results on MSA indicative of staphylococci? (yes or no)

_____

How do you know?

_____

Which isolate (on left or right) is *S. aureus*?

_____

c. Are these results on blood agar indicative of *S. epidermidis* or *S. aureus*?

_____

How do you know?

_____

Isolation and Identification of Streptococci from the Throat

Background

Normal Flora of the Throat

The human throat is a region of the body that has a resident, or normal, bacterial flora. The predominant throat residents are species of streptococci. The most common streptococci are the nonpathogenic viridans streptococci. An opportunistically pathogenic species of streptococci, *Streptococcus pneumoniae*, the causative agent of pneumococcal pneumonia, is found in the throat of 30–70% of normal individuals. It can enter the lungs to cause pneumonia when an individual's resistance is weakened by a primary infection, such as influenza. Other resident members of the throat flora include species of staphylococci, including in some cases *Staphylococcus aureus*, species of *Neisseria*, and diphtheroids.

Pathogens of the Throat

The pathogenic species *Streptococcus pyogenes* is not considered a normal member of the resident throat flora.

However, this organism can enter the throat through the air via aerosol droplets from an infected individual. Once in the throat, it can cause strep throat, a condition characterized by a sore throat, high fever, and a red, inflamed appearance at the back of the throat.

Identification of Throat Isolates

The Gram stain is used to determine the morphology and Gram reaction of throat isolates. The catalase test is used to differentiate the streptococci, which are catalase-negative, from the staphylococci/micrococci, which are catalase-positive. The Gram-positive cocci that are catalase-negative are considered presumptive species of streptococci.

Streptococci common in the throat are differentiated based on the tests listed in table 22.1. For example, certain streptococci produce hemolysins that completely lyse red blood cells, resulting in a clear zone around colonies on blood agar. This reaction, called **beta-hemolysis**, differentiates the beta-hemolytic streptococci from the alpha-hemolytic streptococci. These latter streptococci produce hemolysins that only partially

Table 22.1 Differentiation of Species of Streptococci Commonly Found in the Throat

Strep name	Hemolysis	Bacitracin-susceptible	Optochin-susceptible
Group A (*Streptococcus pyogenes*)	beta	(+)	N/A
Non–group A, beta-hemolytic	beta	(−)	N/A
Streptococcus pneumoniae	alpha	N/A	(+)
Viridans streptococci	alpha	N/A	(−)

lyse red blood cells, producing a green color around colonies on blood agar. This reaction is called **alpha-hemolysis**. The beta-hemolytic streptococci are further differentiated based on their susceptibility to the antibiotic bacitracin. Bacitracin-susceptible, beta-hemolytic streptococci are pathogenic and referred to as group A streptococci, or *Streptococcus pyogenes*. The alpha-hemolytic streptococci are further differentiated based on their susceptibility to ethylhydrocupreine (optochin). Optochin-susceptible, alpha-hemolytic streptococci are considered presumptive *Streptococcus pneumoniae*, while optochin-resistant, alpha-hemolytic streptococci are considered nonpathogenic viridans streptococci.

Materials

Cultures (24–48-hour on blood agar)
Streptococcus pneumoniae
Streptococcus pyogenes

 All agents in red are BSL2 bacteria.

Media
Blood agar plates (tryptic soy agar with 5% sheep blood)

Chemicals and reagents
Bacitracin (A) disks
Gram-stain reagents
Hydrogen peroxide (for catalase test)
Optochin disks

Equipment
Incubator (35°C)
Light microscope

Miscellaneous supplies
Bibulous paper
Biohazard bag (for waste disposal)
Bottle with tap water
Bunsen burner and striker
Cotton-tipped swab, sterile
Disposable gloves
Forceps
Glass microscope slides
Immersion oil
Inoculating loop
Lens paper
Ruler (mm)
Tongue depressor
Wax pencil

Procedure

First Session: Isolation of Bacteria from the Throat

1. After putting on disposable gloves, take a sterile cotton-tipped swab in your right hand and a tongue depressor in your left hand, or vice versa if left-handed.

2. Hold down the tongue of your lab partner with the tongue depressor while moving the cotton-tipped end of the swab toward the back of the throat. **Do not touch any other part of the mouth**. Touch the swab to the back of the throat. Rub the cotton-tipped end over the back of the throat as shown in figure 22.1. Withdraw the swab from the mouth without touching any other surface. Give the swab to your lab partner so that

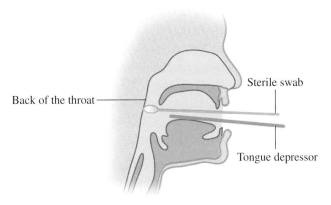

(a)

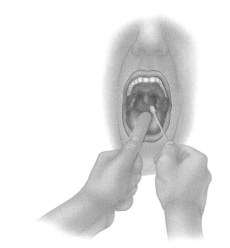

(b)

Figure 22.1 Procedure for obtaining a throat swab. (a) Side view. (b) Front view.

he or she can use it to inoculate a blood agar plate. Discard the tongue depressor in a biohazard bag or similar container.

3. To inoculate a blood agar plate, rub the swab back and forth over one area of the plate near the edge (figure 22.2a). Dispose of the swab in a biohazard bag or similar container. With a sterile loop, cross the swabbed area to spread the bacteria. Repeat this spreading process a second and third time as depicted (figure 22.2c, d). Label your plate. Dispose of the gloves in a biohazard bag or similar container.

4. Place the inoculated plate into a 35°C incubator.

Second Session

Selection of Throat Isolates

1. After 24–48 hours, examine your throat culture plate. Determine the total number of different bacterial types. Record this number in the laboratory report.

2. Examine your plates for signs of hemolysis. Look for a green discoloration, indicative of alpha-hemolysis, and clearing around colonies, indicative of beta-hemolysis. Record the presence of these reactions.

3. Select from the plate one common bacterial type that is alpha-hemolytic, and one common bacterial type that is beta-hemolytic. If there are no beta-hemolytic colonies, then select two alpha-hemolytic types. Number these #1 and #2, and record their colony morphology and hemolytic reaction. Do the same for two known cultures, *Streptococcus pneumoniae* and *S. pyogenes*.

Identification of Throat Isolates

The two common throat isolates will be identified using the tests specified in figure 22.3. You will also test *Streptococcus pneumoniae* and *Streptococcus pyogenes* in conjunction with your unknowns.

1. Gram-stain your unknown throat isolates and the two known cultures. Record your results in the laboratory report.

2. If one or both of your unknowns are Gram-positive cocci, continue your identification by doing a catalase test as follows: Use a sterile loop to place some cells from the unknown culture into a drop of water on a glass slide. Add a drop of hydrogen peroxide. Watch for bubbles, indicative of a positive test. The absence of bubbles indicates a negative test. Do a catalase test for your two known cultures as well. Record your results.

3. If one or both unknowns is a catalase-negative, Gram-positive coccus, then you have isolated presumptive streptococci. If so, continue on to step 4 for your unknown and known cultures.

4. Do a separate streak-plate on blood agar for each culture. Using sterile forceps, place a 0.04 U bacitracin (A) disk in streak quadrants 1 and 2 for beta-hemolytic streptococci. Using sterile forceps, place an optochin disk in streak quadrants 1 and 2 for alpha-hemolytic streptococci. Incubate the plates at 35°C.

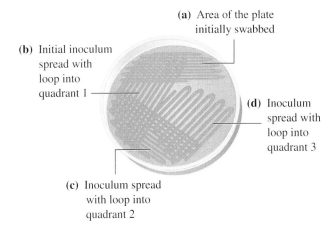

(a) Area of the plate initially swabbed

(b) Initial inoculum spread with loop into quadrant 1

(c) Inoculum spread with loop into quadrant 2

(d) Inoculum spread with loop into quadrant 3

Figure 22.2 Steps in the inoculation of a blood agar plate with a throat culture.

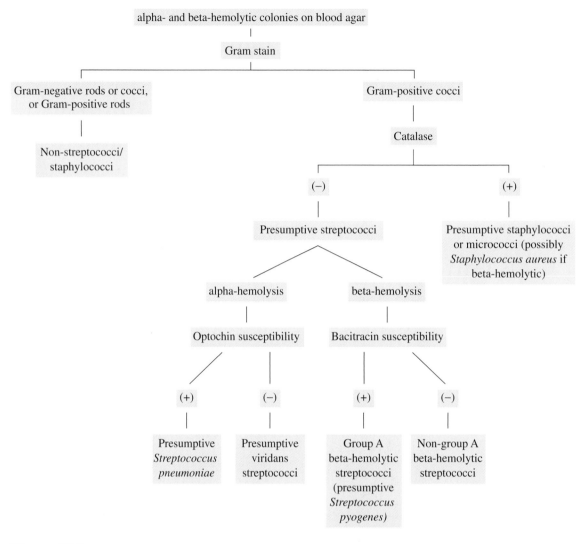

Figure 22.3 Identification scheme for common streptococci from the throat.

Third Session

Identification of Streptococci

1. After 24–48 hours, examine the blood agar plates streaked with beta-hemolytic streptococci. If the growth of beta-hemolytic streptococci is absent around the bacitracin disk, indicating susceptibility, your isolate is *Streptococcus pyogenes*, the causative agent of strep throat. The presence of this organism does not necessarily indicate an active case of strep throat, since it may occur normally in low numbers in some individuals. However, a large number of colonies of this isolate on your original plate (hundreds) indicates an active case of strep throat.

2. After 24–48 hours, examine the blood agar plates streaked with alpha-hemolytic streptococci. If a zone of inhibition greater than or equal to 14 mm occurs around the optochin disk, this indicates susceptibility and the occurrence of *Streptococcus pneumoniae*, the causative agent of pneumococcal pneumonia. The presence of this organism in the throat is considered normal in many individuals, and does not indicate an active case of pneumonia. If there is no zone of inhibition or if the zone around the optochin disk measures less than 14 mm, your isolate is resistant to optochin and a member of the viridans streptococci, a nonpathogenic group that represents the most common form of streptococci in the throat.

LABORATORY REPORT

22

NAME _____ DATE _____

LAB SECTION _____

Isolation and Identification of Streptococci from the Throat

1. Total number of bacterial types from your throat = _____

2. Hemolysis on blood agar plates:

   alpha-hemolysis present? (yes or no) _____

   beta-hemolysis present? (yes or no) _____

3. Throat isolates

   ### a. Identification of streptococci

Hemolytic isolates	Colony morphology	Type of hemolysis (alpha or beta)	Cell morphology and Gram reaction	Catalase	*Results indicative of streptococci?
#1					
#2					
Known cultures: *Streptococcus pneumoniae*					
Streptococcus pyogenes					

*If yes, continue with (b) and (c); if no, stop here.

   ### b. Differentiation of beta-hemolytic streptococci

Beta-hemolytic throat isolate	Bacitracin susceptibility	Identification
Which # or #'s?		
Known culture: *Streptococcus pyogenes*		

   ### c. Differentiation of alpha-hemolytic streptococci

Alpha-hemolytic throat isolate	Optochin susceptibility	Identification
Which # or #'s?		
Known culture: *Streptococcus pneumoniae*		

4. Did you isolate nonpathogenic viridans streptococci from your throat? Is this a common result? Explain.

5. Did you isolate *S. pneumoniae* from your throat? Is this a common result? Explain.

6. Did you isolate *S. pyogenes* from your throat? Is this a common result? Explain.

7. Answer the following questions based on these photographs:

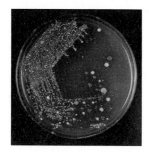

a. Does this individual have strep throat?

   _____

   How do you know? _____

b. Is the colony morphology and hemolysis of this throat isolate consistent with *Streptococcus pyogenes*?_____

   Explain. _____

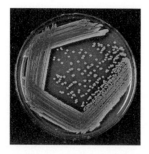

c. This Gram-positive, catalase-positive coccus is beta-hemolytic on blood agar. Does the presence of this isolate in the throat indicate strep throat? _____

   Explain. _____

d. This isolate is a Gram-positive coccus, catalase-negative, and optochin-resistant. Is this isolate part of the normal throat flora? _____

   Explain. _____

A. Identification of Enteric Bacteria, Including the Intestinal Pathogens *Salmonella* and *Shigella*
B. Enterotubes, a Rapid Test System to Identify Enteric Bacteria

Background

Normal Flora of the Intestinal Tract

The large intestine offers an ideal environment for the survival of a large number of resident bacteria. The most common intestinal bacteria are members of the family *Enterobacteriaceae*. These bacteria are often referred to as **enterics** and include normally nonpathogenic species, such as *Escherichia, Enterobacter, Klebsiella, Proteus*, and *Citrobacter*. Non-enteric bacteria are also common in the intestine, including *Enterococcus (Streptococcus) faecalis, Pseudomonas aeruginosa*, and *Staphylococcus aureus*.

Pathogens of the Intestinal Tract

While the nonpathogenic enterics and non-enterics are always present in large numbers, there are several bacteria that invade the intestinal tract after being ingested in contaminated food or water. Among these intestinal invaders are two genera of enteric bacteria, *Salmonella* and *Shigella*. Although they share characteristics with nonpathogenic enterics, these enterics are considered pathogens and not normal residents. *Salmonella*, when ingested, causes the intestinal disease salmonellosis, while *Shigella*, when ingested, causes the intestinal disease called shigellosis, or bacterial dysentery.

Identification of Intestinal Bacteria

Enterics are Gram-negative rods, facultatively anaerobic, and oxidase-negative. A key trait used to differentiate enterics is lactose utilization, which is easily determined on **MacConkey agar** (table 23.1). Enterics that utilize lactose are called lactose fermenters; they turn the medium red and include nonpathogenic enterics such as *Escherichia* and *Enterobacter*. Enterics that do not utilize lactose are called lactose nonfermenters; they produce no color change on MacConkey

Table 23.1 Composition of MacConkey Agar (MAC)

Peptone	17 g
Proteose peptone	3 g
Lactose	10 g
Bile salts	1.5 g
Sodium chloride	5 g
Neutral red	0.03 g
Crystal violet	0.001 g
Agar	13.5 g
Distilled water	1,000 ml
Final pH	7.1

Source: *The Difco Manual*. Eleventh Edition. Difco Laboratories.

agar and include the nonpathogenic enteric *Proteus* and the pathogenic enterics *Salmonella* and *Shigella*. Therefore, lactose fermentation is a critical test for distinguishing nonpathogenic enterics from pathogenic enterics. The medium **triple sugar iron (TSI) agar** (table 23.2) is useful in distinguishing lactose nonfermenters (table 23.3), while MR-VP medium and SIM medium (see Exercise 19) are useful for distinguishing lactose fermenters.

Enterotube, a Rapid Test System to Identify Enteric Bacteria

Recently, rapid test systems have been commercially developed for the identification of enteric bacteria. These systems incorporate a large number of tests into a single unit. All tests are inoculated at once, and the results, which are obtained in 24–48 hours, provide sufficient information for the identification of an isolate.

Table 23.2 Composition of Triple Sugar Iron (TSI) Agar

Beef extract	3 g	Ferrous sulfate	0.2 g
Yeast extract	3 g	Sodium chloride	5 g
Peptone	15 g	Sodium thiosulfate	0.3 g
Proteose peptone	5 g	Phenol red	0.024 g
Dextrose	1 g	Agar	12 g
Lactose	10 g	Distilled water	1,000 ml
Sucrose	10 g	Final pH	7.4

Source: *The Difco Manual.* Eleventh Edition. Difco Laboratories.

Table 23.3 Differentiation of Three Lactose Non-fermenting Enterics Using Triple Sugar Iron Agar

Enteric	Slant color (lactose and/or sucrose fermentation)	Butt color (glucose fermentation)	H₂S production	Designation
Proteus vulgaris	Yellow (+)	Yellow (+)	Black (+)	A/A, H₂S
Salmonella typhimurium	Red (−)	Yellow (+)	Black (+)	Alk/A, H₂S
Shigella flexneri	Red (−)	Yellow (+)	No black (−)	Alk/A

One such test system, depicted in figure 23.1*a*, is called the **Enterotube® II**. It contains 12 compartments in a single unit that accommodates 15 biochemical tests. The compartments are all inoculated at once by pulling an inoculating wire through the unit. In 18–24 hours, the color changes in compartments are noted, and a test is scored as either positive or negative (figure 23.1*b*). Positive tests are used to determine a 5-digit identification number, which identifies the unknown (figure 23.1*c*).

Materials

Cultures (24–48 hour on agar)
  Enterobacter aerogenes, a nonpathogenic enteric
  Enterococcus (Streptococcus) faecalis, a non-enteric
  Escherichia coli, a nonpathogenic enteric
  Proteus vulgaris, a nonpathogenic enteric
  Pseudomonas aeruginosa, a non-enteric
  Salmonella typhimurium, a pathogenic enteric

  Shigella flexneri, a pathogenic enteric
  Staphylococcus aureus, a non-enteric

 All agents in red are BSL2 bacteria.

Media
  Enterotube® II
  MacConkey (MAC) agar plates
  MR-VP medium tubes
  O-F glucose tubes
  SIM medium tubes
  Tryptic soy agar (TSA) plates
  Triple sugar iron (TSI) agar tubes

Chemicals and reagents
  Gram-stain reagents
  Hydrogen peroxide (for catalase test)
  Kovac's reagent (for indole test)
  Methyl red pH indicator (for methyl red test)
  Mineral oil
  Oxidase reagent

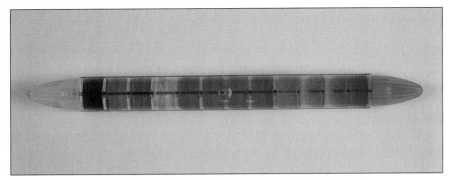

(a) The Enterotube II unit. The 12 compartments are inoculated with the enclosed inoculating wire.

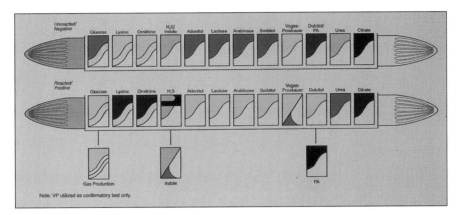

(b) After incubation, each compartment is examined for color change to indicate a positive or negative test result.

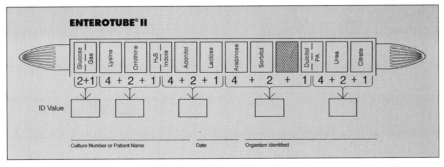

(c) Positive test results are circled to yield a 5-digit ID value used to identify the unknown.

Figure 23.1 An outline of the Enterotube II procedure.

Equipment
   Incubator (35°C)
   Light microscope

Miscellaneous supplies
   Bibulous paper
   Bottle with tap water
   Bunsen burner and striker
   Disposable gloves

Glass microscope slides
Immersion oil
Inoculating loop
Inoculating needle
Lens paper
Pasteur pipette with bulb
Test tube rack
Wax pencil

Procedure

First Session: Identification of Intestinal Bacteria

You will be assigned three unknown cultures to identify: a non-enteric, a nonpathogenic enteric, and a pathogenic enteric. You will first need to determine which of the three are enterics by following the identification scheme in figure 23.2.

1. Do a Gram stain on each of your three unknown cultures. Record your results in the laboratory report. If one of your cultures is a Gram-positive coccus, do a catalase test to complete your identification of this non-enteric (figure 23.2).

2. For Gram-negative rods, inoculate two O-F glucose tubes (cover the medium in one tube with sterile mineral oil) and a tryptic soy agar (TSA) plate. Place the tubes and plate in a 35°C incubator.

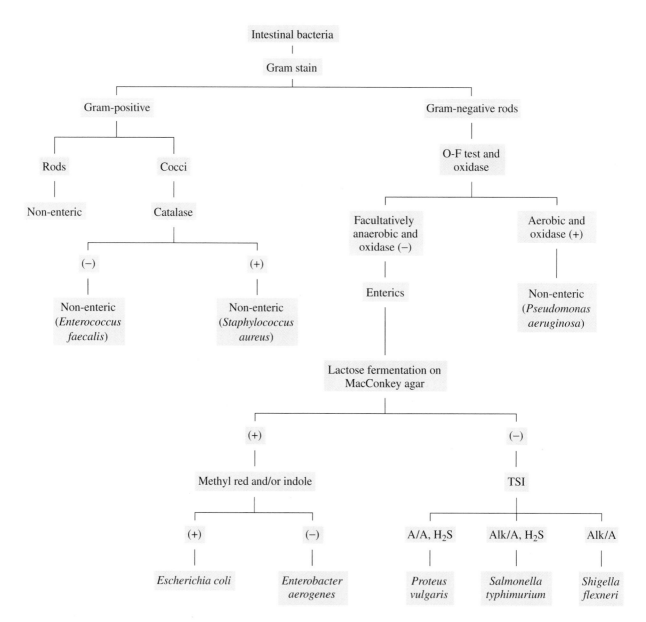

Figure 23.2 Identification scheme for non-enterics, nonpathogenic enterics, and pathogenic enterics.

Second Session: Identification of Enterics

1. After 24–48 hours, examine the O-F glucose tubes for color changes. Enterics are facultatively anaerobic and will turn both tubes yellow. Now place a drop of oxidase reagent on colonies of the TSA plate. Enterics are oxidase-negative, so there should be no color change. Oxidase-positive colonies turn purple. Record your results for both tests.

2. You have now identified your non-enteric, whether Gram-positive or Gram-negative.

3. Use figure 23.2 to identify one of your enterics. For this, inoculate a MacConkey agar plate, and incubate at 35°C.

4. Use the Enterotube® II stepwise procedure outlined in figure 23.3 to identify the other enteric.

Third Session: Differentiation of Enterics

1. After 24–48 hours, examine your MacConkey agar plate for color changes. Lactose fermenters will appear red, while non-fermenters will appear colorless. If positive for lactose fermentation, inoculate an MR-VP medium tube and/or a SIM medium tube. If negative for lactose fermentation, inoculate a TSI agar tube with an inoculating needle. Stab the butt first, and then streak the slant. Incubate inoculated tubes at 35°C.

2. After 18–24 hours, examine the compartments of the Enterotube® II, and record positive and negative tests using the information provided in table 23.4. Use positive test results to calculate your identification number (see example in figure 23.3). Consult the information booklet provided to find the bacterium that matches this code number.

Fourth Session: Differentiation of Enterics (continued)

1. *Lactose fermenters:* After 72 hours, add 10 drops of methyl red pH indicator to your MR-VP medium tube. After mixing, examine for color change (a red color is positive, while a yellow color is negative). Add 5 drops of Kovac's reagent to your SIM medium tube. If the reagent turns red, the culture is indole-positive; if it remains yellow, the culture is indole-negative. Record these results and your identification.

2. *Lactose non-fermenters:* After 24 hours, examine the TSI agar tube. Record the reactions for the slant and butt and whether or not hydrogen sulfide was produced. Refer to table 23.3 for help in interpreting your results. Record your identification.

(a) Remove organisms from a well-islolated colony. Avoid touching the agar with the wire. To prevent damaging Enterotube II media, do not heat-sterilize the inoculating wire.

(b) Inoculate each compartment by first twisting the wire and then withdrawing it all the way out through the 12 compartments, using a turning movement.

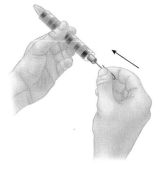

(c) Reinsert the wire (without sterilizing), using a turning motion through all 12 compartments until the notch on the wire is aligned with the opening of the tube.

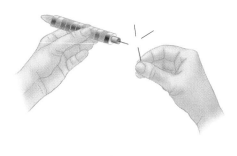

(d) Break the wire at the notch by bending. The portion of the wire remaining in the tube maintains anaerobic conditions essential for true fermentation.

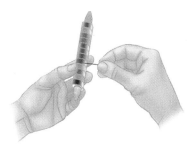

(e) Punch holes with broken-off part of wire through the thin plastic covering over depressions on sides of the last eight compartments (adonitol through citrate). Replace caps, and incubate at 35°C for 18–24 hours.

(f) After interpreting and recording positive results on the sides of the tube, perform the indole test by injecting 1 or 2 drops of Kovac's reagent into the H_2S/indole compartment.

O R N	H_2S	I N D	A D O N	L A C	A R A B	S O R B	D U L	P A	U R E A
+ ② + 1		④ + 2 + ①		4 + ② + ①			④ + ② +

② ⑤ ⬡ ⬡

(g) Perform the Voges-Proskauer test, if needed for confirmation, by injecting the reagents into the H_2S/indole compartment.

   After encircling the numbers of the positive tests on the laboratory report, total up the numbers of each bracketed series to determine the 5-digit code number. Refer to the *Enterotube II Interpretation Guide* for identification of the unknown by using the code number.

Figure 23.3 Steps in the Enterotube II procedure.

Table 23.4 Enterotube II Reactions

Compartment	Medium/Test	Description and Interpretation of Tests	Results Summary Positive	Negative
1	Glucose (GLU)	Tests for glucose fermentation. A shift in pH is indicated by a change in color of the medium from red to yellow, reflecting the production of acidic fermentation by-products. A change in the color of the medium from red to yellow should be interpreted as a positive reaction. Orange should be interpreted as negative.	Yellow	Red/orange
	Gas production (GAS)	Gas from fermentation is indicated as a definite separation of the wax overlay from the surface of the culture medium. Bubbles in the culture medium should not be interpreted as evidence of gas production.	Separation of wax	No separation of wax
2	Lysine decarboxylation (LYS)	Measures the ability of bacteria to decarboxylate lysine to produce the alkaline by-product cadaverine. Any shift in the color of the culture medium from yellow to purple should be interpreted as a positive reaction. The medium should remain yellow if decarboxylation does not take place.	Purple	Yellow
3	Ornithine decarboxylation (ORN)	Measures the ability of bacteria to decarboxylate ornithine to produce the alkaline by-product putresine. Any shift in the color of the culture medium from yellow to purple should be interpreted as a positive reaction. The medium should remain yellow if decarboxylation does not take place.	Purple	Yellow
4	H_2S production (H_2S)	H_2S is produced from the metabolism of sulfur containing compounds (e.g., thiosulfate and amino acids) in the culture medium. Ferrous (Fe^{2+}) ions in the medium react with the H_2S to produce the black precipitate (FeS). Any blackening of the medium indicates that H_2S has been produced.	Black	No change
	Indole formation (IND)	Indole is produced when tryptophan is degraded by the enzyme tryptophanase. After injection of Kovac's reagent into the medium (after 18–24 hours of incubation), any indole present will react with the reagent to produce a pink-red color.	Red	No change
5	Adonitol (ADON)	Tests for adonitol fermentation. A shift in pH is indicated by a change in the color of the medium from red to yellow, reflecting the production of acidic fermentation by-products. A change in the color of the medium from red to yellow should be interpreted as a positive reaction. Orange should be interpreted as negative.	Yellow	Red/ orange
6	Lactose (LAC)	Tests for lactose fermentation. A shift in pH is indicated by a change in the color of the medium from red to yellow, reflecting the production of acidic fermentation by-products. A change in the color of the medium from red to yellow should be interpreted as a positive reaction. Orange should be interpreted as negative.	Yellow	Red/ orange

Table 23.4 Enterotube II Reactions *(continued)*

Compartment	Medium/Test	Description and Interpretation of Tests	Results Summary Positive	Negative
7	Arabinose (ARAB)	Tests for arabinose fermentation. A shift in pH is indicated by a change in the color of the medium from red to yellow, reflecting the production of acidic fermentation by-products. A change in the color of the medium from red to yellow should be interpreted as a positive reaction. Orange should be interpreted as negative.	Yellow	Red/ orange
8	Sorbitol (SORB)	Tests for sorbitol fermentation. A shift in pH is indicated by a change in the color of the medium from red to yellow, reflecting the production of acidic fermentation by-products. A change in the color of the medium from red to yellow should be interpreted as a positive reaction. Orange should be interpreted as negative.	Yellow	Red/ orange
9	Voges-Proskauer (VP)	Tests for the production of acetoin, an intermediate in the 2,3-butanediol fermentation pathway. Acetoin is detected by the injection of 2 drops of solution containing 20% KOH and 0.3% creatine and 3 drops of alpha-napthol solution (5% wt/vol alpha-napthol in absolute ethanol). The development of a pink-red color 10 to 20 minutes after the addition of the alpha-napthol solution indicates that acetoin was produced.	Pink	Colorless
10	Dulcitol (DUL)	Tests for dulcitol fermentation. A shift in pH is indicated by a change in the color of the medium from green to yellow, reflecting the production of acidic fermentation by-products. A change in the color of the medium to yellow or pale yellow should be interpreted as a positive reaction.	Yellow	Green
	Phenylalanine deaminase (PA)	Test for the formation of pyruvic acid from the deamination of phenylalanine. Pyruvic acid reacts with Fe^{3+} in the medium to cause a gray to black discoloration.	Black	Yellow
11	Urea (UREA)	Test for urease production. Hydrolysis of urea results in the production of ammonium, which makes the medium alkaline and causes a color change from yellow to red-purple. Light pink and other shades of red should be interpreted as positive.	Red-violet	Yellow
12	Citrate (CIT)	Test for the ability of certain bacteria to use citrate as the sole source of carbon. Utilization of citrate results in the production of alkaline metabolites, which turn the pH indicator in the culture medium from green to royal blue. Any intensity of blue should be interpreted as positive.	Blue	Green

Source: Becton Dickinson Microbiology Systems, Cockeysville, MD 21030.

LABORATORY REPORT

23

NAME ——————————————————— DATE —————————————

LAB SECTION ————————————————————————

A. Identification of Enteric Bacteria, Including the Intestinal Pathogens *Salmonella* and *Shigella*

B. Enterotubes, A Rapid Test System to Identify Enteric Bacteria

1. Unknown nos. ——————— ——————— ———————

2. Record the results of your tests for the three unknown cultures.

Unknown no.	Cell morphology	Gram reaction	Catalase (if Gram-positive coccus)	O-F test	Oxidase	Enteric?

3. Identify the non-enteric based on your data in the preceding table. —————————————————

4. Identify an enteric using figure 23.2.

Unknown no.	Lactose fermentation on MAC agar	Methyl red	Indole	TSI	Identification

5. Identify an enteric using Enterotube® II.

Unknown no.	Code no. obtained	Identification

6. Answer the following questions based on these photographs:

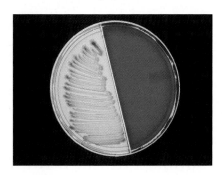

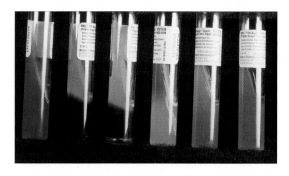

a. The culture on the left is growing on MAC agar. Is the culture a lactose fermenter?

_____

How do you know?

_____

b. Which TSI agar tube is indicative of the results of *Shigella flexneri*?

_____

Isolation and Identification of Bacteria from the Urinary Tract

Background

Normal Flora of the Urinary Tract

The urinary tract is normally sterile, but the urine released from it can become contaminated by bacteria that inhabit the distal end of the urethra and the external genitalia. Even so, the number of bacteria in urine is typically low (i.e., ranging from 0 to 10,000 bacteria/ml). This range is considered normal.

Pathogens of the Urinary Tract

When bacteria invade the urinary tract and cause a **urinary tract infection (UTI),** their presence is reflected in extremely high numbers in urine (i.e., in excess of 100,000 bacteria/ml of urine). Bacteria capable of invading the urinary tract and causing UTIs include *Escherichia coli*, *Enterobacter aerogenes*, *Pseudomonas aeruginosa*, *Proteus vulgaris*, *Enterococcus faecalis*, and *Staphylococcus saprophyticus*.

Identification of Urinary Tract Isolates

A Gram stain is done on an isolate from urine that numbers in excess of 100,000 bacteria/ml. If a Gram-positive coccus is found, a catalase test will determine whether it is a staphylococcus (positive) or streptococcus/enterococcus (negative). If the culture is catalase-negative, the use of BEA, or **bile esculin agar** (table 24.1), will confirm the group D streptococci, since only these bacteria can tolerate the high bile content of this agar while hydrolyzing esculin, a reaction that yields a dark brown color.

A Gram-negative rod can be tested for oxidase. A positive reaction may indicate *Pseudomonas aeruginosa*. A negative oxidase test is indicative of the enteric bacteria, such as *Proteus vulgaris*, *Escherichia coli*, and *Enterobacter aerogenes*. These bacteria can be differentiated by triple sugar iron (TSI) agar (see Exercise 23), the methyl red test (see Exercise 19), and the indole test (see Exercise 19).

Table 24.1 The composition of Bile Esculin Agar (BEA)

Beef extract	3 g
Peptone	5 g
Esculin	1 g
Oxgall	40 g
Ferric citrate	0.5 g
Agar	15 g
Distilled water	1,000 ml
Final pH	6.6

Source: *The Difco Manual*. Eleventh Edition. Difco Laboratories.

Materials

Cultures (on agar plates)
  Enterobacter aerogenes
  Enterococcus faecalis
  Escherichia coli
  Proteus vulgaris
  Pseudomonas aeruginosa
  Staphylococcus saprophyticus

Media
  Bile esculin agar (BEA) tubes
  MR-VP medium tubes
  SIM medium tubes
  Triple sugar iron (TSI) agar tubes
  Tryptic soy agar (TSA) plates

Chemicals and reagents
  Gram-stain reagents
  Hydrogen peroxide (for catalase test)
  Kovac's reagent
  Methyl red pH indicator
  Oxidase reagent

Equipment
  Incubator (35°C)
  Light microscope

Miscellaneous supplies
  Bibulous paper
  Biohazard bag
  Bottle of tap water
  Bunsen burner and striker
  Container and bag (to collect urine)
  Disposable gloves
  Glass microscope slides
  Immersion oil
  Inoculating loop (standard)
  Inoculating loop (5 µl)
  Inoculating needle
  Lens paper
  Pasteur pipette with bulb
  Test tube rack
  Towelette (to clean urethral opening)
  Wax pencil

Procedure

First Session: Collection and Inoculation of Urine

1. Wash your hands. Use an antiseptic towelette to clean around the opening of the urethra.

2. Collect a midstream sample of urine in a clean, plastic container. Put the lid on, and close tightly. Place the container in a plastic bag. Store in the refrigerator if the urine will not be cultured within 1–2 hours.

> **CAUTION:** *This step should be done wearing gloves under a safety hood or behind a plastic shield placed on the countertop.*

3. When ready to culture, mix the urine, and then dip a 5 µl inoculating loop into the fluid. Streak the 5 µl of urine obtained onto a tryptic soy agar (TSA) plate using the method depicted in figure 24.1. Repeat this process for a second plate. Number the plates #1 and #2. Discard the remaining urine in the restroom, and then deposit the gloves, urine container, bag, and loop in a biohazard bag or similar waste container.

4. Place both plates in a 35°C incubator.

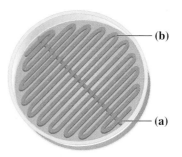

Figure 24.1 Inoculation of a TSA plate with urine. (a) Initial line streak. (b) Back-and-forth streak across initial line streak.

Second Session: Isolation and Identification of Urinary Tract Bacteria

1. After 48–72 hours, examine your culture plates. Count the number of bacterial colonies on each plate, average this number, and then use this average to calculate the number of bacteria per milliliter of urine. (To do this, multiply the average number of bacterial colonies by a factor of 200.) Record your results in the laboratory report.

2. If any of the bacteria on your plate exceeded 100,000 per milliliter of urine, you may have a UTI caused by this organism. Identify this organism and one assigned unknown culture using the scheme depicted in figure 24.2. If you do not have a UTI, then you will identify two assigned unknown cultures using this scheme.

3. Do a Gram stain of your two cultures. Record their morphology and Gram reaction. For Gram-positive cocci, do a catalase test. If catalase-positive, the culture is *Staphylococcus*. If catalase-negative, inoculate a BEA tube and place it in a 35°C incubator. For Gram-negative rods, do an oxidase test. If oxidase-positive, the culture is *Pseudomonas*. If oxidase-negative, inoculate a TSI agar tube, a MR-VP medium tube, and/or a SIM medium tube. Place these tubes in a 35°C incubator.

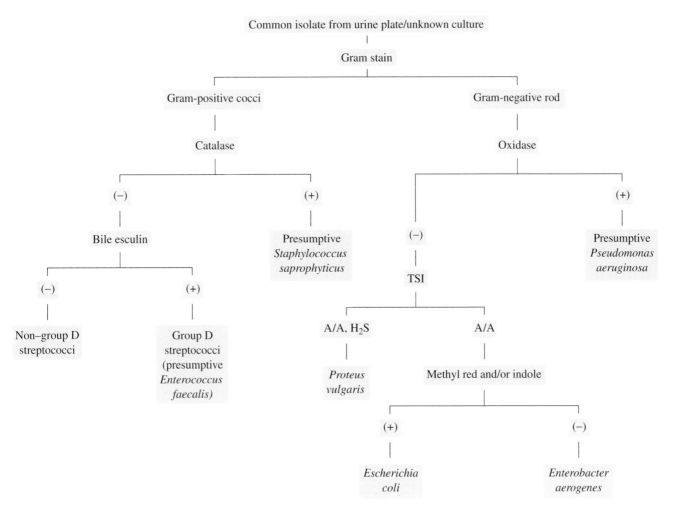

Figure 24.2 Identification scheme for bacteria that commonly cause urinary tract infections.

Third Session: Identification of Urinary Tract Bacteria (continued)

1. After 24–48 hours, examine your BEA slant for the presence of a dark brown color, indicative of group D streptococci. The absence of a dark brown color indicates a non–group D streptococcus.

2. After 24–48 hours, examine the TSI agar tube. A yellow butt and slant with a black discoloration is indicative of *Proteus vulgaris*. If there is no black discoloration, examine your MR-VP medium tube using step 3 and/or your SIM medium tube using step 4.

3. After 72 hours, add 10 drops of methyl red pH indicator to the MR-VP medium tube, and mix. A red color indicates a positive methyl red test, while a yellow color indicates a negative methyl red test.

4. After 24–48 hours, add 5 drops of Kovac's reagent to your SIM medium tube, and watch for the development of a red color in the reagent, a positive indole test. No color change is a negative indole test.

LABORATORY REPORT

24

NAME _____ DATE _____

LAB SECTION _____

Isolation and Identification of Bacteria from the Urinary Tract

1. Record the number of bacteria in the urine sample.

Plate	Number of bacteria/5 µl of urine
#1	
#2	

Average = _____ bacteria/5 µl $\times$ 200 = _____ bacteria/ml

2. Did the average number of bacteria/ml urine fall within the normal range of 0–10,000 bacteria/ml? If yes, how do you account for these bacteria in urine?

3. Did the average number of bacteria/ml urine exceed 100,000 bacteria/ml? If yes, what does this number indicate?

4. Identify the urinary tract bacteria in your sample.

Test	UTI isolate or unknown #1	Unknown #2
Cell morphology		
Gram reaction		
Catalase		
BEA		
Oxidase		
TSI		
Methyl red		
Indole		

UTI isolate or unknown #1   Unknown #2

Identification: _____   _____

5. Answer the following questions based on these photographs:

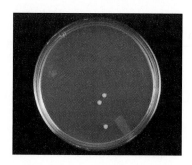

a. This TSA plate was inoculated with

   5 µl of urine. Is a UTI indicated? _____

   Explain. _____

   _____

b. This organism, responsible for a UTI,
   gave the above reaction on BEA. What
   organism is indicated?

   _____

Assessing Antibiotic Effectiveness: The Kirby-Bauer Method

Background

Among the various chemical agents used to control microbial growth, **antibiotics** are unique because they are selective in their action—that is, they specifically target bacterial cells. For this reason, they can be introduced into the human body to treat disease with minimal effects on human cells. Since Alexander Fleming discovered penicillin produced by a mold in his laboratory over 60 years ago, antibiotics have become a standard method used by physicians to treat bacterial diseases.

Types of Antibiotics

Since the discovery of penicillin, many other useful antibiotics have been developed. Each antibiotic has a specific mechanism of action against bacteria. In some cases, the action is **broad-spectrum**, or effective against a wide variety of bacteria. In others, the action is **narrow-spectrum**, or effective against only certain bacteria. Table 25.1 lists 10 selected antibiotics, their effect on cells, and their spectrum of activity.

Table 25.1 Antibiotics Used to Treat Bacterial Infections

Antibiotic	Effect on cells	Spectrum of activity
Ampicillin	Inhibits cell wall synthesis	Broad-spectrum antibiotic [effective against Gram (+) cocci and some Gram (−) bacteria]
Bacitracin	Inhibits cell wall synthesis	Narrow-spectrum antibiotic [effective against Gram (+) bacteria]
Chloramphenicol	Inhibits protein synthesis	Broad-spectrum antibiotic [effective against Gram (+) and Gram (−) bacteria]
Erythromycin	Inhibits protein synthesis	Narrow-spectrum antibiotic [effective against Gram (+) bacteria]
Gentamicin	Inhibits protein synthesis	Broad-spectrum antibiotic [effective against Gram (+) and Gram (−) bacteria]
Penicillin G	Inhibits cell wall synthesis	Narrow-spectrum antibiotic [effective against Gram (+) cocci]
Polymyxin B	Disrupts cell membrane	Narrow-spectrum antibiotic [effective against Gram (−) rods]
Streptomycin	Inhibits protein synthesis	Broad-spectrum antibiotic [effective against Gram (+) and Gram (−) bacteria]
Tetracycline	Inhibits protein synthesis	Broad-spectrum antibiotic [effective against Gram (+) and Gram (−) bacteria]
Vancomycin	Inhibits cell wall synthesis	Narrow-spectrum antibiotic [effective against Gram (+) bacteria]

Evaluating Effectiveness

When a disease-causing bacterium is isolated from a patient, the physician must determine which antibiotic to administer for treatment. The most widely used method to evaluate the effectiveness of antibiotics against specific bacteria is the **Kirby-Bauer method**. In this method, outlined in figure 25.1, **Mueller-Hinton agar** (table 25.2) is inoculated with a culture of a bacterial isolate. After inoculation, antibiotic disks are placed on the agar surface. Plates are incubated to allow for bacterial growth and then inspected for **zones of inhibition** around antibiotic disks. Zones of inhibition are measured in millimeters and compared to an interpretive standard to determine if the isolate is susceptible or resistant to the antibiotic. Antibiotics that the organism is susceptible to are candidates for use in treating the patient.

Table 25.2	Composition of Mueller-Hinton Agar
Beef infusion	300 g
Casamino acids	17.5 g
Starch	1.5 g
Agar	17 g
Distilled water	1,000 ml
Final pH	7.3

Source: *The Difco Manual.* Eleventh Edition. Difco Laboratories.

Materials

Cultures (24 hours in tryptic soy broth)
  Bacillus cereus, Gram-positive rod
  Escherichia coli, Gram-negative rod
  Pseudomonas aeruginosa, Gram-negative rod
  Staphylococcus aureus, Gram-positive coccus

 SAFETY *All agents in red are BSL2 bacteria.*

Media
  Mueller-Hinton agar plates, 4 mm thick
    (25 ml/plate)

Chemicals and reagents
  Antibiotic disks (loose in sterile petri dish)
    Ampicillin
    Bacitracin
    Chloramphenicol
    Erythromycin
    Gentamicin
    Penicillin G
    Polymyxin B
    Streptomycin
    Tetracycline
    Vancomycin
  Ethanol, 70%

Equipment
  Incubator (35°C)

Miscellaneous supplies
  Beaker, 250 ml
  Bunsen burner and striker
  Cotton-tipped swabs, sterile
  Disposable gloves
  Forceps
  Ruler, millimeter
  Test tube rack
  Wax pencil

(a) Dip a cotton-tipped swab into a broth culture.

(b) Spread the culture over the entire plate. Dip and spread two more times, as depicted in figure 25.2.

(c) Sterilize forceps by dipping the end in alcohol and then flaming.

(d) Pick up antibiotic disk, and place on inoculated plate. Repeat for four other antibiotics.

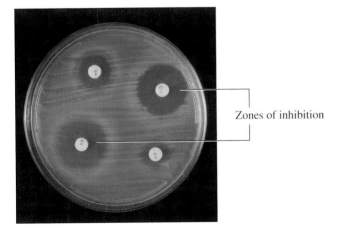

Zones of inhibition

(e) After incubation, examine plates for zones of inhibition, indicative of antibiotic effectiveness.

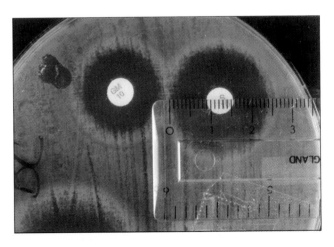

(f) Measure zones of inhibition to the nearest millimeter (mm), and compare to the interpretive standard. This zone measures 24 mm.

Figure 25.1 An outline of the Kirby-Bauer method.

Procedure

First Session: Preparation of Plates

1. Dip a sterile cotton-tipped swab into one of the broth cultures, and use it to inoculate a Mueller-Hinton agar plate using the procedure depicted in figure 25.2. Inoculation of the plate in this way ensures a lawn of bacterial growth after incubation. Repeat this inoculation procedure for a second plate using the same organism. Label both plates.

2. Repeat step 1 for the other three cultures. You should now have a total of eight inoculated and labeled plates, two for each culture. After inoculation, allow all plates to dry for 15 minutes before proceeding to the next step.

3. Pour some 70% ethanol into a 250 ml beaker.

CAUTION *Caution: Keep this beaker away from your flame.*

   a. Dip your forceps into the alcohol, and then pass the forceps over the Bunsen burner flame to sterilize them.

   b. Now pick up an antibiotic disk from one of the petri dishes, and place it on one of your inoculated plates.

   c. After placement on the agar, tap it once to make sure it is secure.

Repeat steps a–c until you have placed this disk on a plate for each culture. Proceed to the next disk until five disks have been placed on a plate for each culture. Place the other five disks on the second plate, for a total of 10 disks per culture.

4. When all disks are in place, put your eight plates into a 35°C incubator.

Second Session: Examination of Plates

1. Your plates must be examined after 16–18 hours of incubation. If you cannot examine them then, place them in a refrigerator until examination.

2. Examine your plates for zones of inhibition. Measure these with a millimeter ruler across the disk as shown in figure 25.1*f*. Record the

(a) Area of initial swab   **(b)** Area of second swab

(c) Area of third swab

Figure 25.2 Inoculation of a Mueller-Hinton agar plate. (a) Dip a cotton swab in the culture, and swab across the surface of the agar without leaving any gaps. (b) Using the same swab, dip in the culture again, and swab the agar in a direction perpendicular to the first inoculum. (c) Dip and swab a third time at a 45° angle to the first inoculum.

diameter of the zone to the nearest whole millimeter in the laboratory report. If only one side of the zone can be measured, multiply the number obtained by 2 to obtain a full zone of inhibition. If there is no zone (i.e., if growth occurs up to the edge of the disk), record a zero. *Note:* You might see colonies within the zone of inhibition. These colonies consist of cells that are resistant to the antibiotic. Continue recording the zones of inhibition until you have all 40 measurements.

3. Now compare the zone of inhibition you obtained to the interpretive standards for these antibiotics in table 25.3. Record whether each organism is resistant, susceptible, or intermediate to the antibiotic.

4. Complete this exercise by recording for each type of bacteria the antibiotics the organism is susceptible to. These represent possible drugs of choice to treat infections by these bacteria.

Table 25.3 Interpretive Standards for Antibiotics Selected for This Exercise

Antimicrobial agent	Abbreviation	Concentration	Diameter of zone of inhibition (mm)		
			Resistant	*Intermediate*	*Susceptible*
Ampicillin	AM	10 µg	—	—	—
Gram-negative	—	—	11	12–13	14
Staphylococci	—	—	20	21–28	29
Bacitracin	B	10 units	8	9–12	13
Chloramphenicol	C	30 µg	12	13–17	18
Erythromycin	E	15 µg	13	14–22	23
Gentamicin	GM	10 µg	12	13–14	15
Penicillin G	P	10 units	—	—	—
Staphylococci	—	—	20	21–28	29
Other organisms	—	—	11	12–21	22
Polymyxin B	PB	300 units	8	9–11	12
Streptomycin	S	10 µg	11	12–14	15
Tetracycline	TE	30 µg	14	15–18	19
Vancomycin	VA	30 µg	9	10-11	12

Source: *Antimicrobial Susceptibility Test Discs*. Technical information published by Becton Dickinson Microbiology Systems, Cockeysville, Maryland.

LABORATORY REPORT

25

NAME _____ DATE _____

LAB SECTION _____

Assessing Antibiotic Effectiveness: The Kirby-Bauer Method

1 Record the diameter of the zones of inhibition (in mm).

Antibiotic	Disk code	Bacteria			
		Bacillus cereus	*Escherichia coli*	*Pseudomonas aeruginosa*	*Staphylococcus aureus*
Ampicillin	AM10				
Bacitracin	B10				
Chloramphenicol	C30				
Erythromycin	E15				
Gentamicin	GM10				
Penicillin G	P10				
Polymyxin B	PB300				
Streptomycin	S10				
Tetracycline	TE30				
Vancomycin	VA30				

2. Evaluate the bacteria based on the interpretive standards in table 25.3. Record whether each type is R = resistant, S = susceptible, or I = intermediate based on the standard.

Antibiotic	Disk code	Bacteria			
		Bacillus cereus	*Escherichia coli*	*Pseudomonas aeruginosa*	*Staphylococcus aureus*
Ampicillin	AM10				
Bacitracin	B10				
Chloramphenicol	C30				
Erythromycin	E15				
Gentamicin	GM10				
Penicillin G	P10				
Polymyxin B	PB300				
Streptomycin	S10				
Tetracycline	TE30				
Vancomycin	VA30				

3. Which antibiotics are the test organisms susceptible to, and hence, candidates for treating infections caused by these organisms?

Bacteria	Antibiotics susceptible to (candidates for treatment)
Bacillus cereus	
Escherichia coli	
Pseudomonas aeruginosa	
Staphylococcus aureus	

4. In your results, did you find evidence of broad-spectrum antibiotics (i.e., ones effective against both Gram-positive and Gram-negative bacteria)? Which antibiotics were broad-spectrum? Were these results as you expected?

5. In your results, did you find evidence of narrow-spectrum antibiotics (i.e., ones effective against only either Gram-positive or Gram-negative bacteria)? Which antibiotics were narrow-spectrum? Were these results as you expected?

6. Answer the following questions based on these photographs:

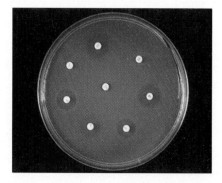

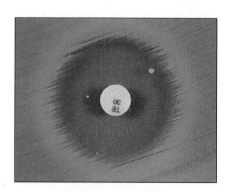

a. Name the clear area around some of these antibiotic disks. _____

What do they indicate?

_____

b. Explain these two colonies within the clear area around this TE 30 disk. _____

_____

Identification of a Clinical Bacterial Unknown

Background

Previous exercises have covered aspects of clinical isolates from various regions of the body, such as the skin, throat, intestinal tract, and urinary tract. In this exercise, you will apply what you have learned about clinical isolate diagnosis to the identification of a clinical bacterial unknown.

Materials

Cultures (24–48-hour broth)
  Enterobacter aerogenes
  Enterococcus faecalis
  Escherichia coli
  Klebsiella pneumoniae
  Proteus vulgaris
  Pseudomonas aeruginosa
  Salmonella typhimurium
  Shigella flexneri
  Staphylococcus aureus
  Staphylococcus epidermidis
  Streptococcus pneumoniae
  Streptococcus pyogenes

 All agents in red are BSL2 bacteria.

Stains
  Gram-stain reagents

Media
  Bile esculin agar (BEA) tubes
  Blood agar plates
  Lactose broth tubes
  Mannitol salt agar (MSA) plates
  Motility test agar tubes
  MR-VP medium tubes
  O-F glucose tubes
  SIM medium tubes
  Simmons citrate agar tubes
  Tryptic soy agar (TSA) plates
  Urea broth tubes

Chemicals and reagents
  Bacitracin (A) disks
  Coagulase test (rabbit plasma or rapid latex agglutination test kit)
  Hydrogen peroxide (catalase test)
  Methyl red pH indicator
  Optochin disks
  Oxidase reagent

Equipment
  Incubator (35°C)

Miscellaneous supplies
  Bottle with tap water
  Bunsen burner and striker
  Disposable gloves
  Glass slides
  Immersion oil
  Inoculating loop and needle
  Lens paper
  Mineral oil (sterile)
  Pasteur pipette with bulb
  Staining tray
  Test tube rack
  Wax pencil

Procedure

1. You will select an unknown culture, or one will be assigned to you. In either case, record the number of your unknown in the laboratory report.

2. Begin with a streak-plate on TSA. After incubation, examine the streak-plate to make sure you have good growth and a pure culture. If growth is slow or poor, try a streak-plate on a blood agar plate.

3. After recording culture characteristics, do a Gram stain (see Exercise 11) on your streak-plate culture to determine cell shape, cell arrangement, and Gram reaction. Record your results.

4. Examine the identification scheme in figure 26.1 to determine the test to be done next. If a catalase test is required, see Exercise 19; for coagulase

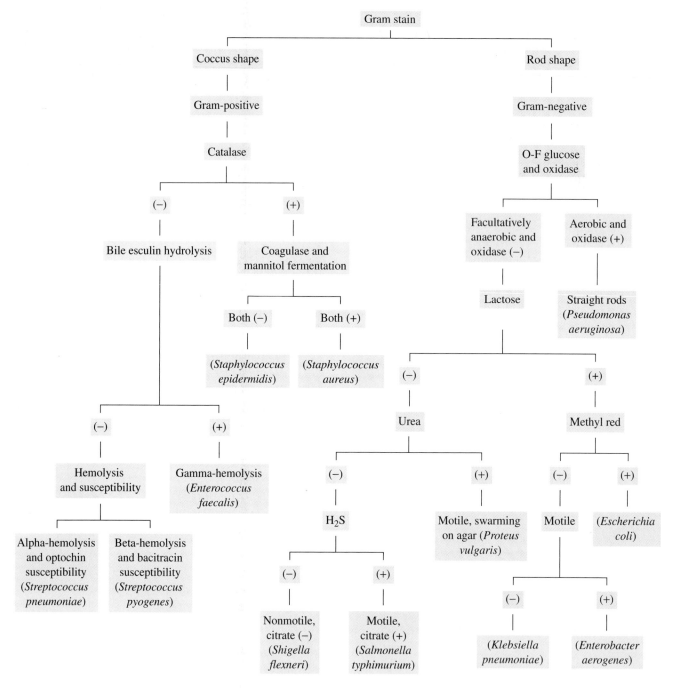

Figure 26.1 Identification scheme for 12 clinical bacterial unknowns.

and mannitol fermentation tests, see Exercise 21; for bile esculin hydrolysis, see Exercise 24; for hemolysis and susceptibility tests, see Exercise 22; for O-F glucose, oxidase, lactose, methyl red, and H_2S tests, see Exercise 19; for motility, see Exercise 18.

5. A urea test is required to separate rapid urea utilizers, such as *Proteus vulgaris*, from other bacteria. The test employs **urea broth** (table 26.1) and is read 18–24 hours after inoculation. Urea utilization turns the medium pink; nonutilization displays no color change.

Table 26.1 Composition of Urea Broth

Yeast extract	0.1 g
Potassium phosphate	18.6 g
Urea	20 g
Phenol red	0.01 g
Distilled water	1,000 ml
Final pH	6.8

Source: *The Difco Manual.* Eleventh Edition. Difco Laboratories.

6. If citrate utilization is required, inoculate a tube of **Simmons citrate agar** (table 26.2) and examine for color change. A blue color after 24–48 hours incubation indicates a citrate-positive test; no color change is a negative test.

7. Continue with your tests until your unknown has been identified. Be sure to record the results of all tests and the identity of your unknown in the laboratory report.

Table 26.2 Composition of Simmons Citrate Agar

Magnesium sulfate	0.2 g
Ammonium dihydrogen phosphate	1 g
Dipotassium phosphate	1 g
Sodium citrate	2 g
Sodium chloride	5 g
Agar	15 g
Bromthymol blue	0.08 g
Distilled water	1,000 ml
Final pH	6.8

Source: *The Difco Manual.* Eleventh Edition. Difco Laboratories.

8. Your laboratory instructor may wish to see all results when you are finished. Therefore, keep all slides, plates, and tubes until examined by your laboratory instructor.

LABORATORY REPORT

NAME _____ DATE _____

LAB SECTION _____

Identification of a Clinical Bacterial Unknown

Unknown no. _____

1. Follow the identification scheme in figure 26.1 to identify your clinical bacterial unknown. Be sure to perform only the tests required to identify your unknown.

2. Record your results for the required tests.

Procedure	Observations	Results
Culture characteristics *Broth*		
Agar		
Staining characteristics *Cell shape*		
Cell arrangement		
Gram reaction		
Biochemical/other characteristics *Bacitracin susceptibility*		
Bile esculin hydrolysis		
Catalase test		
Citrate utilization		
Coagulase		
Hemolysis		
Hydrogen sulfide (H_2S) production		
Lactose utilization		
Mannitol fermentation		
Methyl red test		
Motility		
O-F glucose test		
Optochin susceptibility		
Oxidase test		
Urea utilization		

3. After following the scheme in figure 26.1 and recording the results for the required tests, I conclude that my unknown is _____.

4. Answer the following questions based on these photographs:

a. This culture is growing in urea broth.

    Is the culture urea-positive? _____

    How do you know?

    _____

b. These two cultures are growing on citrate agar. Does the culture on the right utilize

    citrate? _____

    How do you know?

    _____

Controlling the Risk and Spread of Bacterial Infections

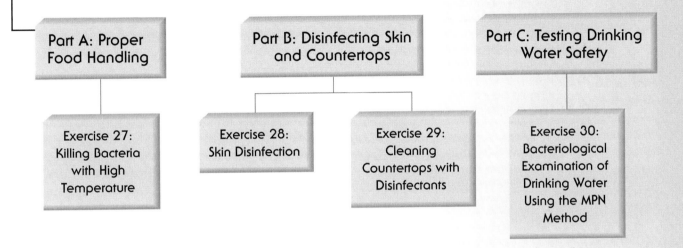

Part A: Proper Food Handling

Exercise 27: Killing Bacteria with High Temperature

Part B: Disinfecting Skin and Countertops

Exercise 28: Skin Disinfection

Exercise 29: Cleaning Countertops with Disinfectants

Part C: Testing Drinking Water Safety

Exercise 30: Bacteriological Examination of Drinking Water Using the MPN Method

27

Killing Bacteria with High Temperature

Background

Dry and Wet (Moist) Heat

Heat is one of the most effective methods used to kill bacteria. Heat is generally divided into dry and wet (moist) heat (table 27.1). Dry heat, which includes incineration and the hot-air oven, kills bacteria by oxidizing components of the cell. Wet (moist) heat, which includes boiling water, autoclave/pressure cooker, pasteurization, and fractional sterilization, kills bacteria by coagulating proteins in the cell, including essential enzymes and cell structures.

Using Dry Heat in the Kitchen

Dry heat is used for grilling on the stovetop or baking in the oven. When properly used, dry heat in the kitchen can effectively eliminate the risk of contracting certain types of bacterial diseases.

Pathogenic strains of *Escherichia coli*, such as the 0157:H7 strain, cause diarrhea, and can be contracted by eating undercooked hamburger. Cooking hamburger meat to a temperature of 80°C or above should kill all vegeta-tive cells of *E. coli*, if present. Likewise, species of *Salmonella*, such as *S. enteritidis* and *S. typhimurium*, are associated with eating undercooked chicken and eggs, causing salmonellosis. The thorough grilling or baking of chicken and eggs to a temperature of 80°C or above should kill all vegetative cells of *Salmonella*, if present.

Using Wet Heat in the Kitchen

Boiling water has been used for a long time around the home in cooking and disinfecting items, such as baby bottles and canning jars. Drinking water may also require boiling on occasion. For example, whenever water flow is interrupted in water lines by a rupture or drop in pressure, there is a chance of bacterial contaminants entering the water supply. In these cases, city officials may advise people to boil their water prior to use. This eliminates the risk of contracting a waterborne infection until normal service is restored.

In summary, when properly used, heat is an effective household tool to eliminate the risk of bacterial infection. This exercise will demonstrate the killing power of wet heat.

Table 27.1 Types of Heat Used to Kill Bacteria

Type of heat	Examples	Effect on cells	Uses
Dry	Incineration	Oxidizes cell components	Used to sterilize laboratory loops and needles; used to destroy waste and infectious materials
	Hot-air oven	Oxidizes cell components	Used to sterilize laboratory glassware; used in home cooking
Wet	Boiling water	Coagulates cell proteins	Used in home disinfection and cooking
	Autoclave/pressure cooker	Coagulates cell proteins	Autoclave used to sterilize laboratory media; pressure cooker used in home cooking/canning
	Pasteurization	Coagulates cell proteins	Used to disinfect liquids (e.g., milk) to increase shelf life and kill pathogens
	Fractional sterilization	Coagulates cell proteins	Used to sterilize heat-sensitive instruments and chemicals

Materials

Culture (24-hour in tryptic soy broth)
Escherichia coli

Media
Tryptic soy broth tubes (18): 16 × 150 mm
tubes containing 5 ml broth per tube, capped

Equipment
Incubator (35°C)

Miscellaneous supplies
Beaker (1 liter)
Bunsen burner and striker
Pipette (1 ml, sterile); pipette bulb
Test tube rack
Thermometer (°C)
Tripod with ceramic-lined wire mesh
Wax pencil

Procedure

First Session: Inoculation and Heating of Broth Tubes

1. Place a pipette bulb onto a 1 ml sterile pipette and fill the pipette with the broth culture of *E. coli*.

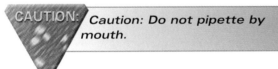

Caution: Do not pipette by mouth.

Figure 27.1 Experimental setup for heating broth tubes inoculated with *Escherichia coli.*

This should be sufficient culture to inoculate 17 of the 18 broth tubes.

2. Aseptically transfer 1 drop of culture to each of 17 broth tubes. *Note:* Insert the pipette into the tube close to the surface of the liquid, and aim the drop directly into the liquid. A drop deposited on the side of the glass may not reach the broth, resulting in a false negative.

3. Thoroughly mix the drop into the broth. Place one of the inoculated tubes in a test tube rack. Label this tube the control. Place the remaining 16 inoculated tubes in the 1 liter beaker, and fill the beaker with tap water to a level above the broth. Now carefully insert the thermometer in the uninoculated broth tube, and place the tube in the water.

4. Place the beaker on the wire mesh platform mounted on the tripod. Move a lighted Bunsen burner to a position beneath the tripod to heat the water. Examine figure 27.1 to see this experimental setup without the 16 inoculated tubes.

5. During heating, remove one tube at every 5°C interval, beginning at 25°C. Label each tube with the temperature at which it was removed, and place it in the test tube rack with the control tube. When the water reaches 100°C, remove the last tube, and turn off the Bunsen burner.

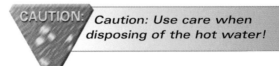

Caution: Use care when disposing of the hot water!

6. Place the test tube rack with the 17 tubes in a 35°C incubator.

Second Session: Examination of Broth Tubes

1. After 48 hours, examine each tube for growth. If viable cells remained after heating, they will have multiplied into millions of cells, turning the broth cloudy or turbid. In this case, you will not be able to see through the liquid. Score these tubes as (+) for growth, indicating that the temperature wasn't sufficient to kill all vegetative cells. If all vegetative cells were killed after heating, none will have been left to multiply, leaving the broth clear. In this case, you will be able to see through the liquid. Score these tubes as (−) for growth, indicating that the temperature was sufficient to kill all vegetative cells. Record your score for each tube in the laboratory report.

2. Continue scoring tubes as (+) or (−) using the criteria in step 1 until all tubes have been scored. Evaluate the results of your experiment as related to the use of heat in your home.

LABORATORY REPORT

NAME _____ DATE _____

LAB SECTION _____

Killing Bacteria with High Temperature

1. In the following table, record your scores for each tube; use a (+) for tubes with cloudy, turbid growth; use a (−) for tubes with clear broth.

Temperature (°C)	Broth turbid (T) or clear (C)?	Growth (+) or (−)?	Heat killed all vegetative cells?
25			
30			
35			
40			
45			
50			
55			
60			
65			
70			
75			
80			
85			
90			
95			
100			

2. According to your results in this experiment, what is the minimum temperature required to kill all vegetative cells of *E. coli*? What application might this have for cooking your hamburger meat at home?

3. If you received a notice from city officials to boil your water before use, would boiling kill *E. coli* and other vegetative bacterial cells if they were present? Explain.

Skin Disinfection: Evaluating Antiseptics and Hand Sanitizers

Background

A variety of chemical agents display antimicrobial activity against bacteria. One category of antimicrobial chemical agents, the antibiotics, was examined in Exercise 25. Two other categories of chemical agents commonly used in the household are antiseptics and disinfectants. **Antiseptics** are chemicals safe enough to be applied to the skin; they are used to prevent wound infections and to disinfect skin. Some commonly used antiseptics and their effects on bacterial cells are presented in table 28.1.

The effectiveness of these skin-applied chemical agents will be examined in this exercise. **Disinfectants** are chemicals considered too harsh to be applied to the skin, and are only used on inanimate surfaces. Disinfectants will be evaluated in Exercise 29.

Evaluating Antiseptics: The Filter Paper Method

Antiseptics are commonly used on the skin to prevent wound infections. One of the ways to determine the effectiveness of antiseptics is to use the **filter paper method**, outlined in figure 28.1. In this method, filter paper disks are dipped into an antiseptic and then placed on an agar plate that has been inoculated with a bacterial culture. The plate is then incubated to allow bacterial growth. After growth, plates are examined for **zones of inhibition** around the chemical-soaked disks, indicating chemical effectiveness. In this exercise, you will use the filter paper method to examine the effectiveness of antiseptics commonly applied to the skin.

Evaluating Hand Sanitizers

Bacteria are numerous on the hands, and represent both members of the normal flora and transients picked up from the environment. While the normal flora is typically not harmful, transients can be disease-causing agents. One of the simplest and most effective ways to eliminate these transient disease-causing agents is to wash your hands. Hungarian physician Ignaz Semmelweis advocated hand washing as a means of preventing disease transmission in the mid-1800s. This simple task is still recommended today by health-care specialists as one of the most effective means of preventing infection.

Table 28.1 Commonly Used Antiseptics

Chemical agent	Effect on cells	Commercial uses
Alcohol (ethyl or isopropyl)	Dehydrates the cell; alters cell membrane; denatures cell proteins	Skin cleansing and degerming agent; skin antiseptic
Benzalkonium chloride	Alters cell membrane	Skin antiseptics
Cetylpyridinium chloride	Alters cell membrane	Mouthwashes
Hexachlorophene	Alters cell membrane; denatures cell proteins	Soaps and skin antiseptics
Hydrogen peroxide	Oxidizes cell components	Skin antiseptic
Mercurochrome or Merthiolate	Denatures cell proteins	Skin antiseptic
Tincture of iodine	Denatures cell proteins	Skin antiseptic
Triclosan	Alters cell membrane; denatures cell proteins	Antibacterial soaps

(a) Obtain a sterile disk using sterile forceps, and dip the disk halfway into antiseptic to allow the disk to soak up the chemical.

(b) Place the chemical-soaked disk on an inoculated plate. Repeat for three other antiseptics.

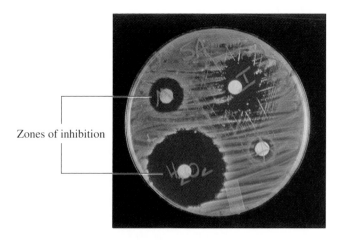

Zones of inhibition

(c) After incubation, examine plates for zones of inhibition, indicative of antiseptic effectiveness.

Figure 28.1 The filter paper method for evaluating antiseptics.

Today, using a hand sanitizer is a popular way to clean the hands. These products are popular because they can be used to disinfect the hands while away from home or when soap, water, or towels are not available. These gel products are dispensed from plastic bottles onto the hands. The hands are then rubbed together until dry. The active ingredient in these products is 62% ethyl alcohol.

This exercise will also evaluate the effectiveness of hand sanitizers in removing bacteria from the hands.

Materials

Cultures (24-hour in tryptic soy broth)
Bacillus cereus
Escherichia coli

Pseudomonas aeruginosa
Staphylococcus aureus

 All agents in red are BSL2 bacteria.

Media
Tryptic soy agar (TSA) plates
Tryptic soy broth tubes

Chemicals and reagents
Antiseptics
Alcohol, ethyl or isopropyl
Benzalkonium chloride (found in skin antiseptics)

Cetylpyridinium chloride (found
in mouthwashes)
Hexachlorophene (found in soaps
and skin antiseptics)
Hydrogen peroxide
Mercurochrome or Merthiolate
Tincture of iodine
Triclosan (found in antibacterial hand soaps)
Ethanol, 70%
Hand sanitizer (active ingredient,
62% ethyl alcohol)

Equipment
Incubator (35°C)

Miscellaneous supplies
Beaker, 250 ml
Bunsen burner and striker
Cotton-tipped swabs, sterile
Filter paper disks, sterile, in a petri dish
Forceps
Wax pencil

Procedure

First Session

Evaluating Antiseptics: The Filter Paper Method

1. Dip a cotton-tipped swab into one of the four cultures, and use it to inoculate a tryptic soy agar plate using the procedure outlined in Exercise 25 (see figure 25.2). *Note:* A lawn of bacterial growth is necessary for this method, as it was for antibiotic testing in Exercise 25. Repeat this inoculation procedure for a second plate using the same culture. Label each plate with a wax pencil.

2. Repeat step 1 for the remaining three cultures. You should now have a total of eight plates, two for each culture. After inoculation, allow all plates to dry for 15 minutes before proceeding to the next step.

3. Pour some 70% ethanol into a 250 ml beaker.

> CAUTION: **Caution: Keep the alcohol away from the flame!**

a. Dip your forceps into the alcohol, and pass them over a Bunsen burner flame to sterilize them.

b. Now pick up a sterile disk with the forceps, and insert it halfway into a drop of the antiseptic poured into a beaker or a petri dish. Let the disk soak up the chemical; when thoroughly soaked, lift the disk and place it on an inoculated plate.

c. After placement, tap the disk lightly to make sure it is secure.

Repeat steps a–c until you have placed this antiseptic on a plate for each culture. Proceed to the next antiseptic until you have placed four disks on a plate for each culture. Place the remaining four antiseptics on the second plate, for a total of eight antiseptics per culture. *Note:* Place the disks as far apart as possible, and mark the antiseptic on the bottom of the plate.

4. When all disks are in place, put your plates into a 35°C incubator.

Evaluating Hand Sanitizers

1. Dip a cotton-tipped swab into a tube of tryptic soy broth to wet the cotton. Rub lightly on the inside of the tube to remove excess liquid.

2. Swab the left hand as follows: Begin at the top of the first finger (nearest the thumb) and swab down to the base of the thumb; roll the swab, and come back up to the fingertip; repeat this two more times to cover this area of the finger and palm (figure 28.2). Use this swab to inoculate a tryptic soy agar plate. Swab the entire surface of the plate, turn 90°, and swab the entire surface again. Be sure to rotate the swab as you go to deposit all the bacteria lifted from the hand. Label this plate "Before, Replicate 1."

3. Repeat step 2 for the third finger of the left hand, swabbing the finger and palm as before with a fresh swab, and then transferring the bacteria lifted to a second tryptic soy agar plate. Label this plate "Before, Replicate 2."

4. Take the hand sanitizer, and place a thumbnail-sized amount in the palm of the left hand. Rub the palms of both hands together, covering all inside surfaces of the hands with sanitizer. Continue rubbing until the gel has disappeared and the hands are dry.

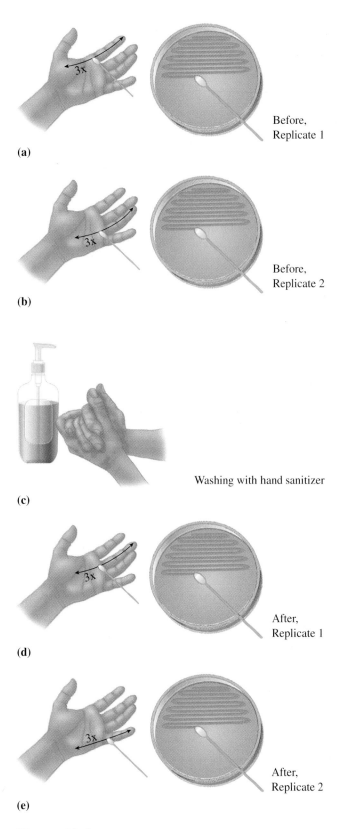

(a)

Before, Replicate 1

(b)

Before, Replicate 2

(c)

Washing with hand sanitizer

(d)

After, Replicate 1

(e)

After, Replicate 2

Figure 28.2 Testing the effectiveness of hand sanitizers.

5. After sanitizer treatment, take a fresh swab, and wet it in broth as before. Swab the second finger, starting at the tip and moving downward to the base of the palm. Rotate the swab, and move upward to the fingertip. Repeat this down-and-up process two more times as before (figure 28.2). Inoculate a third tryptic soy agar plate as before. Label this plate "After, Replicate 1."

6. Using a fresh swab, repeat the swabbing procedure in step 5 for the fourth finger (smallest). Inoculate a fourth tryptic soy agar plate as before, and label it "After, Replicate 2."

7. Place these four plates in a 35°C incubator with the antiseptic plates.

Second Session

Examining Antiseptic Plates

1. After 48–72 hours, examine the culture plates containing antiseptic disks. Examine the growth around the disks.

2. For each disk, look for a zone of inhibition. As for antibiotics, these areas indicate the effectiveness of a chemical agent in preventing growth. However, in this case, the diameter of the zone may not equate to a degree of effectiveness, since chemicals vary in their volatility and diffusion through the agar. Therefore, record only a (+) for a zone of inhibition around a disk indicating susceptibility. Record a (−) for no zone of inhibition, indicating resistance.

3. Complete your observation of all disks for the four cultures, recording a (+) or (−) in the laboratory report.

Examining Hand Sanitizer Plates

1. After 48–72 hours, examine the plates inoculated with the swabs of your left hand. Separate these into "before" and "after" plates.

2. Count the total number of colonies on the two replicate "before" plates and the total number of colonies on the two replicate "after" plates. Record these numbers in your laboratory report. Calculate a "before" average and an "after" average.

3. Record the percentage of bacteria killed by the hand sanitizer.

LABORATORY REPORT

28

NAME ———————————————— DATE ————————————

LAB SECTION ————————————————————————————

Skin Disinfection: Evaluating Antiseptics and Hand Sanitizers

Antiseptics

1. In the following table, record your results for antiseptic plates. Record a (+) for the presence of a zone of inhibition around the disk. Record a (−) for no zone of inhibition.

Antiseptic	Culture			
	Bacillus cereus	*Escherichia coli*	*Pseudomonas aeruginosa*	*Staphylococcus aureus*
Benzalkonium chloride				
Cetylpyridinium chloride				
Ethanol (70%)				
Hexachlorophene				
Hydrogen peroxide				
Isopropyl alcohol				
Mercurochrome or Merthiolate				
Tincture of iodine				
Triclosan				

2. Which antiseptic(s), if any, had the widest spectrum of activity? How would this trait make this a useful antiseptic? Explain.

Hand Sanitizer

1. In the following table, record your results for the hand sanitizer. Record the total number of colonies on the two "before" plates and the total number of colonies on the two "after" plates.

	Total number of colonies	
Replicate	*Before hand sanitizer*	*After hand sanitizer*
1		
2		
Average		

2. Calculate the average percent reduction of bacteria on the hand: _____%

3. Did the hand sanitizer remove the large majority of bacteria from your hand? Based on these results, would you buy this product for use when away from home? When would it be useful?

Cleaning Countertops with Disinfectants

Background

Antimicrobial chemical agents are important in the control of microorganisms. Exercise 25 examined the effectiveness of antibiotics, while Exercise 28 evaluated the effectiveness of antiseptics. A third category of chemical agents, **disinfectants**, are considered too harsh for use on or in the human body; however, they are useful on inanimate surfaces. Some of the chemical agents commonly used in disinfectants are listed in table 29.1.

Disinfectants are widely used around the house to remove bacteria from surfaces. Surfaces that require disinfection at home include the kitchen sink and countertops, bathroom sink and countertops, toilet, shower, and bathtub. Similar surfaces that require periodic disinfection are also found in public facilities and at work. Keeping these surfaces clean and low in bacterial numbers is one of the most effective means of controlling the occurrence and spread of infectious agents.

In this exercise, you will evaluate the effectiveness of several commercially available disinfectants containing the chemical compounds listed in table 29.1.

Materials

Media
  Tryptic soy agar plates
  Tryptic soy broth tubes

Chemicals and reagents
  Disinfectants, commercially available (those listed in table 29.1 or others that contain the same chemicals)

Equipment
  Incubator (35°C)

Miscellaneous supplies
  Adhesive tape
  Bottles, spray-dispenser type
  Cotton-tipped swabs, sterile
  Paper towels
  Ruler, metric
  Wax pencil

Table 29.1 Chemical Agents Commonly Used in Disinfectants

Chemical agent	Effect on cells	Commercial uses
Sodium hypochlorite	Oxidizes cell components	Surface disinfectants and bleach
Orthophenylphenol	Denatures cell proteins	Surface disinfectants, such as Lysol
Alkyldimethylbenzyl ammonium chloride	Alters cell membrane	Surface disinfectants, such as Formula 409
Pine oil	Alters cell membrane; denatures cell proteins	Surface disinfectants, such as Pine-Sol

Procedure

First Session: Inoculation of Plates

1. Select two surfaces to be cleaned. A laboratory countertop and a bathroom or kitchen countertop are recommended. If a bathroom or kitchen is unavailable, select a second laboratory countertop.

2. Mark off a 3,600 cm² area of the first surface to be cleaned. Use four 60 cm pieces of adhesive tape to mark the edges of this area. Also place a piece of adhesive tape in the center of this area. The center piece of tape will help delineate four areas within the 3,600 cm² area: an upper left area, upper right area, lower left area, and lower right area. Designate these four areas as test areas for disinfectants 1, 2, 3, and 4, respectively (figure 29.1).

3. In each test area, use pieces of adhesive tape 10 cm long to mark the edges of two adjacent 100 cm² areas, one designated A, before cleaning with disinfectant, and the other designated B, after cleaning with disinfectant (figure 29.1).

Countertop area (3,600 cm²)

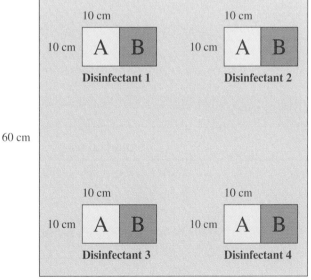

A: Before cleaning, swab each A area with a wet, cotton-tipped swab. Inoculate a tryptic soy agar plate.

B: After cleaning with disinfectant, swab each B area with another swab. Inoculate a second tryptic soy agar plate.

Figure 29.1 Procedure for testing the effectiveness of disinfectants.

4. Dip a sterile, cotton-tipped swab into tryptic soy broth. Use it to swab the 100 cm² area denoted as A, Disinfectant 1. Swab the entire 100 cm² area twice, the second time at a 90° angle to the first. Use the swab to inoculate a tryptic soy agar plate. Rub the swab over the entire surface of the plate, rolling the swab as you do so. Rotate the plate 90° and swab again. Label this plate "A, Disinfectant 1."

5. Take disinfectant 1, and clean the entire disinfectant 1 test area. **Do not spray into any of the other areas.** Prepare the disinfectant per the directions on the container, mixing the disinfectant with water in a spray-type dispenser. In this way, the disinfectant can be thoroughly sprayed over the entire surface before wiping with a paper towel. Be sure to wipe the surface dry. **Do not wipe into any of the other areas.**

6. Dip a fresh cotton-tipped swab in sterile broth, and use it to swab the 100 cm² area denoted as B, Disinfectant 1. Again, be sure to swab the entire 100 cm² area twice. Use this swab to inoculate a second tryptic soy agar plate as before. Label this plate "B, Disinfectant 1."

7. Repeat steps 4–6 to complete the sampling of each A and B area for disinfectants 2, 3, and 4. When finished, you should have inoculated a total of eight tryptic soy agar plates.

8. After completing your sampling of the first surface, repeat steps 2–7 for the second surface. You should have inoculated another eight tryptic soy agar plates for this surface, giving you a total of 16 plates for the two surfaces.

9. Place all plates into a 35°C incubator.

Second Session: Examination of Plates

1. After 48–72 hours, examine your plates. Sort the plates by surface cleaned, disinfectant used, and before cleaning (A) and after cleaning (B). Count the total number of bacterial colonies on each plate, and fill in your results in the laboratory report.

2. Calculate the percent decrease in the bacteria on each cleaned surface for each disinfectant.

LABORATORY REPORT

NAME _____ DATE _____

LAB SECTION _____

Cleaning Countertops with Disinfectants

1. Record the number of colonies on your plates.

    a. **Laboratory countertop (first surface)**

Disinfectant	Before (A)	After (B)	Percent Decrease
1 =			
2 =			
3 =			
4 =			

    b. **Second surface**

Disinfectant	Before (A)	After (B)	Percent Decrease
1 =			
2 =			
3 =			
4 =			

2. Explain the difference between disinfection and sterilization. Which of these terms applies to the action of the chemicals used in this exercise?

3. Do these chemical agents work effectively to remove bacteria from surfaces? Were there any that seemed to work best?

4. Based on your results, do you think the use of these chemicals around the home is justified? If so, when and where would you use these products?

Bacteriological Examination of Drinking Water Using the MPN Method

Background

Coliforms, Indicators of Fecal Contamination

Water is routinely tested to ensure that it is safe for drinking. A widely used indicator of the suitability of drinking water is coliform bacteria. **Coliforms** are Gram-negative, non-endospore-forming rods that are facultatively anaerobic and produce acid and gas from lactose within 48 hours at 35°C. The key indicator organism in this group is *Escherichia coli,* which is not normally present in soil and water, but present in large numbers in the intestines and feces, and capable of long-term survival in the environment. Therefore, the presence of *E. coli* is indicative of human or animal fecal waste. Water contaminated with fecal material, as determined by the presence of coliforms, is considered **nonpotable**, meaning unsuitable for drinking. Water that is coliform-free is considered **potable** and safe for drinking.

Human fecal waste may also carry intestinal pathogens, such as *Salmonella typhi,* the cause of typhoid fever; *Salmonella typhimurium,* the cause of salmonellosis; *Vibrio cholerae,* the cause of cholera; and *Shigella sonnei,* the cause of shigellosis. Each of these intestinal pathogens is transmitted by fecal contamination of drinking water. However, their presence is difficult to detect since they do not typically occur in large numbers and do not survive long in soil and water. As a consequence, coliforms, especially *E. coli,* are used as the indicator of fecal contamination.

Testing Water for Coliforms

One of the methods used to detect coliforms in drinking water is the **most probable number (MPN) method**. This method, outlined in figure 30.1, consists of three parts: (1) a presumptive test; (2) a confirmed test; and (3) a completed test.

In the **presumptive test**, three series of five tubes each, or 15 tubes total, are inoculated with a water sample. Each tube contains 10 ml of lactose broth and a durham tube. Each tube in the first series of five tubes

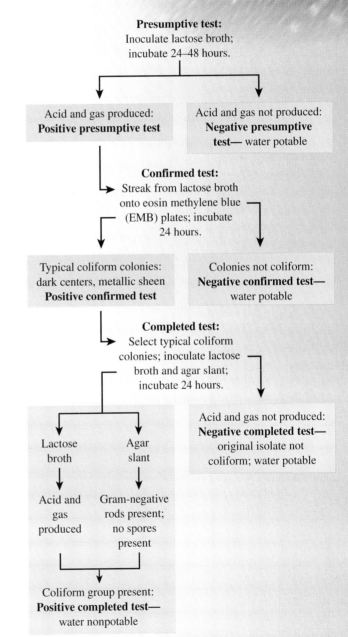

Figure 30.1 The MPN method used to detect coliforms in drinking water.

receives 10 ml of sample; each tube in the second series of five tubes receives 1 ml of sample; and each tube in the third series of five tubes receives 0.1 ml of sample. After 24 hours incubation at 35°C, tubes are examined for the presence of acid and gas, products of lactose fermentation. A positive tube, which has turned yellow and has a gas bubble in the durham tube, is depicted in figure 30.2a. Also depicted is a negative tube, which is unchanged in color and has no gas bubble in the durham tube (figure 30.2b). After 48 hours of incubation, negative tubes are examined again for a delayed positive reaction. All tubes after 48 hours are denoted as either (+) or (−), and a most probable number is assigned according to the index shown in table 30.1. If only one tube scores positive, this is considered a positive presumptive test-that is, it presumes that coliforms are present. However, their presence must be confirmed in the next part. If all tubes score negative, this is considered a negative presumptive test. In this case, the water is considered free of coliforms and, therefore, potable.

In the **confirmed test**, all positive tubes from the highest dilution of sample are streaked onto **eosin methylene blue (EMB) agar** (table 30.2). This agar selects for and differentiates coliform bacteria. *E. coli* is especially easy to differentiate since it produces a distinctive green, metallic sheen on this agar. The presence of colonies on EMB with this characteristic is considered a positive confirmed test—that is, it confirms the presence of coliforms. However, their presence must be further substantiated by the completed test described next. The absence of colonies on EMB with this characteristic is considered a negative confirmed test, and the water is considered absent of coliforms and potable.

In the **completed test**, colonies from EMB with a green, metallic sheen are transferred to a lactose broth tube and a nutrient agar slant. If acid and gas are produced in the lactose broth tube within 24 hours and a Gram stain detects a Gram-negative rod, this is considered a positive completed test, meaning that the confirmation of coliforms in the water is complete. The water is considered contaminated with coliforms and unsafe to drink.

In this exercise, you will use the MPN method to examine the bacteriological quality of three water samples: sewage, surface water, and tap water.

Materials

Water samples
  Sewage

 SAFETY — *Sewage may contain pathogens*

  Surface water (from pond, lake, or stream)
  Tap water

Media
  Eosin methylene blue (EMB) plates
  Lactose broth tubes: each with 10 ml broth and a durham tube, both double-strength and single-strength
  Nutrient agar slant

Chemicals and reagents
  Gram-stain reagents

Equipment
  Incubator (35°C)
  Light microscope

Miscellaneous supplies
  Bunsen burner and striker
  Inoculating loop
  Immersion oil
  Lens paper
  Microscope slides
  Pipettes, 10 ml and 1 ml, sterile; pipette bulb
  Test tube racks
  Wax pencil

Procedure

First Session: Inoculation of Lactose Broth Tubes

1. Take 15 lactose tubes, five double-strength and 10 single-strength, and align into three rows of five in a test tube rack. Place the five double-strength

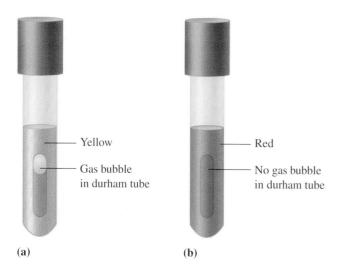

(a)   (b)

Figure 30.2 Lactose broth. (a) Positive tube. (b) Negative tube.

Yellow — (a)
Gas bubble in durham tube — (a)
Red — (b)
No gas bubble in durham tube — (b)

tubes in the front row. In a similar manner, arrange 15 tubes for each of the other two samples, for a total of 45 tubes. Number the tubes in each row 1 to 5; also designate the sample type and sample amount added: 10 ml (front row), 1 ml (middle row), or 0.1 ml (back row).

2. Place a pipette bulb onto a 10 ml pipette.

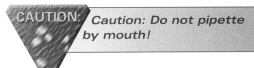

CAUTION: *Caution: Do not pipette by mouth!*

Table 30.1 MPN Index and 95% Confidence Limits for Various Combinations of Positive Results When Five Tubes Are Used per Dilution (10 ml, 1.0 ml, 0.1 ml)

Combination of positives	MPN index/ 100 ml	95% confidence limits Lower	95% confidence limits Upper	Combination of positives	MPN index/ 100 ml	95% confidence limits Lower	95% confidence limits Upper
0-0-0	<2	—	—	4-3-0	27	12	67
0-0-1	3	1.0	10	4-3-1	33	15	77
0-1-0	3	1.0	10	4-4-0	34	16	80
0-2-0	4	1.0	13	5-0-0	23	9.0	86
1-0-0	2	1.0	11	5-0-1	30	10	110
1-0-1	4	1.0	15	5-0-2	40	20	140
1-1-0	4	1.0	15	5-1-0	30	10	120
1-1-1	6	2.0	18	5-1-1	50	10	150
1-2-0	6	2.0	18	5-1-2	60	30	180
2-0-0	4	1.0	17	5-2-0	50	20	170
2-0-1	7	2.0	20	5-2-1	70	30	210
2-1-0	7	2.0	21	5-2-2	90	40	250
2-1-1	9	3.0	24	5-3-0	80	30	250
2-2-0	9	3.0	25	5-3-1	110	40	300
2-3-0	12	5.0	29	5-3-2	140	60	360
3-0-0	8	3.0	24	5-3-3	170	80	410
3-0-1	11	4.0	29	5-4-0	130	50	390
3-1-0	11	4.0	29	5-4-1	170	70	480
3-1-1	14	6.0	35	5-4-2	220	100	580
3-2-0	14	6.0	35	5-4-3	280	120	690
3-2-3	17	7.0	40	5-4-4	350	160	820
4-0-0	13	5.0	38	5-5-0	240	100	940
4-0-1	17	7.0	45	5-5-1	300	100	1,300
4-1-0	17	7.0	46	5-5-2	500	200	2,000
4-1-1	21	9.0	55	5-5-3	900	300	2,900
4-1-2	26	12	63	5-5-4	1,600	600	5,300
4-2-0	22	9.0	56	5-5-5	≥ 1,600	—	—
4-2-1	26	12	65				

Source: *Standard Methods for the Examination of Water and Wastewater*. 18th edition. Copyright 1992 by the American Public Health Association, the American Water Works Association, and the Water Environment Federation. Reprinted with permission.

Table 30.2 Composition of Eosin Methylene Blue (EMB) Agar	
Peptone	10 g
Lactose	5 g
Sucrose	5 g
Dipotassium phosphate	2 g
Eosin Y	0.4 g
Methylene blue	0.065 g
Agar	13.5 g
Distilled water	1,000 ml
Final pH	7.2

Source: *The Difco Manual.* Eleventh Edition. Difco Laboratories.

Add 10 ml of the first sample to each of the five tubes in the front row. Do the same for the second and third samples. Use a fresh 10 ml pipette for each sample.

3. After all the tubes in the front row have been inoculated, use a 1 ml pipette with bulb to inoculate the second and third row of tubes for each sample.

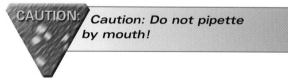

CAUTION *Caution: Do not pipette by mouth!*

The tubes in the second row each receive 1 ml of sample, while those in the third row each receive 0.1 ml. Be sure to change pipettes between each sample.

Place all pipettes that were used on the sewage sample in a disinfectant solution or in some other waste container designated by your laboratory instructor.

4. After completing the inoculation of all tubes, place the test tube racks in a 35°C incubator.

Second Session: Examination of Lactose Broth Tubes (Presumptive Test)

1. After 24–48 hours, examine each tube for the presence of acid and gas. Record tubes with a yellow color and gas as (+) in the laboratory report. Record tubes without a color change or gas as (–). Use the (+) and (–) results to calculate an MPN for each sample (table 30.1).

2. For samples with a positive presumptive test (i.e., one or more tubes with a yellow color and gas), continue to the confirmed test by streak–plating positive tubes of the highest dilution onto EMB agar plates. Place these plates in a 35°C incubator.

Third Session: Examination of EMB Agar Plates (Confirmed Test)

1. After 24–48 hours, examine each EMB plate for the presence of colonies with a green, metallic sheen. The presence of these colonies represents a positive confirmed test, while their absence represents a negative confirmed test.

2. If one or more samples have coliform colonies, continue to the completed test by selecting a green, metallic sheen colony from an EMB plate and using it to inoculate a lactose broth tube and a nutrient agar slant. Place these in a 35°C incubator.

Fourth Session: Examination of Lactose Broth Tube and Gram Stain (Completed Test)

1. After 24 hours, examine the lactose broth tube for acid and gas. If positive, do a Gram stain from the nutrient agar slant to determine if the culture is a Gram-negative rod. If lactose-positive and a Gram-negative rod, the confirmation of coliforms in the sample is complete.

2. Based on your results, determine the potability of each water sample.

LABORATORY REPORT

NAME ———————————————— DATE ——————————

LAB SECTION ————————————————————————

Bacteriological Examination of Drinking Water Using the MPN Method

1. Results for water sample #1: ——————————

    a. **Presumptive test**

    (+) or (−)

Sample added	Tube 1	Tube 2	Tube 3	Tube 4	Tube 5	Number of positive tubes
10 ml						
1 ml						
0.1 ml						

    Combination of positives = ——————————

    MPN index/100 ml = ——————————

    Presumptive test: positive or negative? ——————————

    b. **Confirmed test**

    Number of tubes of highest dilution streaked onto EMB plates ——————————

    Number of these plates with green, metallic-sheen colonies ——————————

    Confirmed test: positive or negative? ——————————

    c. **Completed test**

    Number of green, metallic-sheen colonies selected from EMB plates ——————————

    Number of these colonies that produced acid and gas from lactose and were Gram-negative rods

    ——————————

    Completed test: positive or negative? ——————————

    d. **Conclusion:** Water potable or nonpotable? ——————————

2. Results for water sample #2:_____

   a. **Presumptive test**

   (+) or (−)

Sample added	Tube 1	Tube 2	Tube 3	Tube 4	Tube 5	Number of positive tubes
10 ml						
1 ml						
0.1 ml						

   Combination of positives = _____

   MPN index/100 ml = _____

   Presumptive test: positive or negative? _____

   b. **Confirmed test**

   Number of tubes of highest dilution streaked onto EMB plates _____

   Number of these plates with green, metallic-sheen colonies _____

   Confirmed test: positive or negative? _____

   c. **Completed test**

   Number of green, metallic-sheen colonies selected from EMB plates _____

   Number of these colonies that produced acid and gas from lactose and were Gram-negative rods

   _____

   Completed test: positive or negative? _____

   d. **Conclusion:** Water potable or nonpotable? _____

3. Results for water sample #3: _____

   a. **Presumptive test**

   (+) or (−)

Sample added	Tube 1	Tube 2	Tube 3	Tube 4	Tube 5	Number of positive tubes
10 ml						
1 ml						
0.1 ml						

   Combination of positives = _____

   MPN index/100 ml = _____

   Presumptive test: positive or negative? _____

b. **Confirmed test**

Number of tubes of highest dilution streaked onto EMB plates _____

Number of these plates with green, metallic-sheen colonies _____

Confirmed test: positive or negative? _____

c. **Completed test**

Number of green, metallic-sheen colonies selected from EMB plates _____

Number of these colonies that produced acid and gas from lactose and were Gram-negative rods

_____

Completed test: positive or negative? _____

d. **Conclusion:** Water potable or nonpotable? _____

4. What are coliforms? Why is their presence in drinking water routinely monitored?

5. What action should be taken if coliforms are detected in drinking water?

6. Answer the following questions based on these photographs:

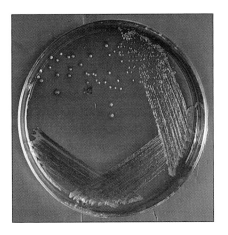

A water sample yielded these results for the presumptive test (left) and the confirmed test (right).

Collectively, what do these results indicate? _____

_____

What would be the next step? _____

Bacterial Genetics

Part A: Bacterial Genomic DNA

Part B: Plasmid DNA

Exercise 31:
Bacterial DNA
Isolation and
Southern Analysis

Exercise 32:
Mutagenesis in
Bacteria:
The Ames Test

Exercise 33:
Plasmid Isolation and
Restriction Mapping

Part C: The Transfer of Drug Resistance

Exercise 34:
Acquiring Antibiotic
Resistance Through
Bacterial Transformation

Exercise 35:
Acquiring Antibiotic
Resistance Through
Bacterial Conjugation

31

Bacterial DNA Isolation and Southern Analysis

Background

The sequence of the genome of one strain of *Escherichia coli*, K12, was completed in 1997 by researchers at the University of Wisconsin, Madison. The genome, con- sisting of a single, circular, double-stranded DNA chro- mosome, is 4,639,221 base pairs long and contains 4,403 genes. A partial genetic map of the *E. coli* K12 chromosome is shown in figure 31.1.

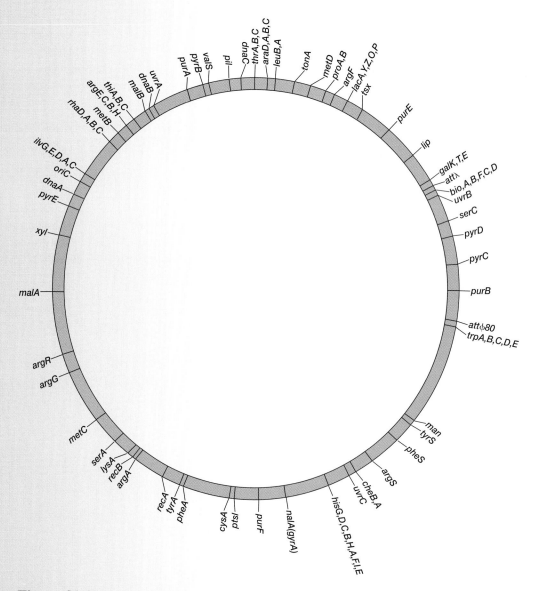

Figure 31.1 Genetic map of *E. coli* K12 with the locations of selected genes. *E. coli* K12 strains are used for fundamental work in biochemistry, genetics, and biotechnology, acting as carriers of genes encoding therapeutic proteins.

In preparation for analysis, the DNA must be isolated from a pure culture of the bacteria. The isolation involves lysing the cells, degrading cellular RNA and protein with enzymes, and separating cellular debris from the DNA through extraction with an organic solvent. The DNA is then cut into fragments with a **restriction endonuclease**, an enzyme that cuts through double-stranded DNA at a particular recognition sequence, (see also Exercise 33 and table 33.1). The restriction enzyme EcoRI, for example, cuts DNA wherever it contains the sequence,

-GAATTC-
-CTTAAG-

Therefore, cutting a series of DNA samples from the same source with EcoRI will always generate the same set of *restriction fragments*. These fragments can be separated by size using gel electrophoresis.

However, cellular DNAs are so long (here, over 4 million base pairs) that when they are cut with a restriction enzyme and the fragments are separated on a typical electrophoresis gel, no clear *restriction pattern* can be seen. Only a smear of DNA representing fragments of just about every possible size is visible (figure 31.2). Think of this DNA smear as a ladder that has so many rungs so close together that you cannot distinguish one rung from the next, or as a barcode that is solid black—there is no information there. Southern blotting allows the detection of a discrete region of the DNA, revealing a restriction pattern of just that part of the genome (figure 31.3). Southern blotting is also often employed to generate DNA fingerprints (see Exercise 36).

In this exercise, you will isolate DNA from bacteria for restriction analysis (figure 31.3 *a–c*). If time permits, you may proceed with a Southern blot over the next few lab sessions (figure 31.3 *d–i*) in order to identify the restriction pattern of the bacterial gene *lacZ*. The *lacZ* gene encodes the enzyme β-galactosidase.

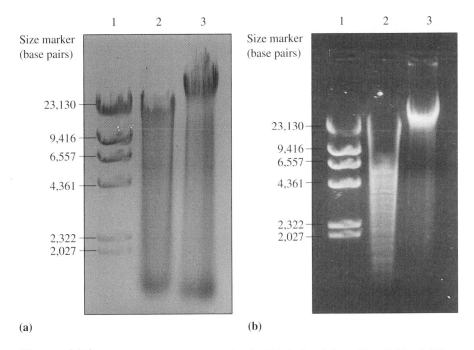

(a)   (b)

Figure 31.2 Agarose gel electrophoresis of DNA isolated from *E. coli*. The 0.8% agarose gels have been stained with (a) methylene blue or (b) ethidium bromide. Both gels contain the following samples: bacteriophage lambda DNA cut with the restriction enzyme HindIII (size marker, lane 1), *E. coli* DNA cut with the restriction enzyme EcoRI (lane 2), and *E. coli* DNA that has not been cut with a restriction enzyme (lane 3). The fragments (bands) in lane 1 are distinct because the lambda genome is only about 49,000 base pairs long, and the enzyme cut the DNA into discernible fragments. The *E. coli* DNA restriction fragment lengths in lane 2 are indistinguishable from one another by this method, and appear as a smear.

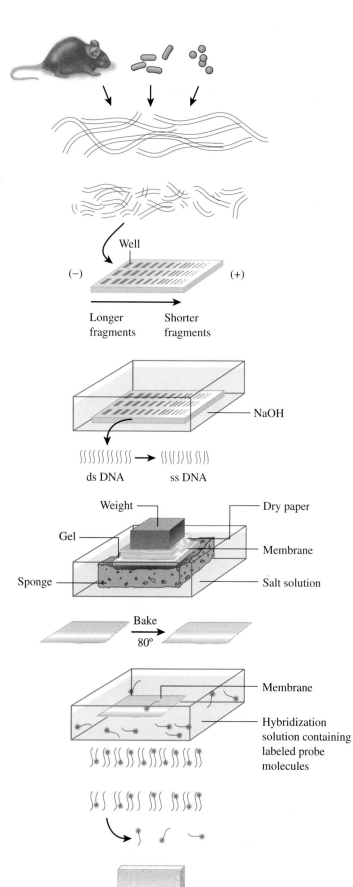

(a) **Isolation of DNA from tissues, cells, or viruses:** The DNA is mechanically sheared during this procedure, generating large fragments.

(b) **Restriction enzyme digestion:** The large fragments of DNA are cut at specific sites with a restriction enzyme, generating restriction fragments characteristic of the organism.

(c) **Agarose gel electrophoresis:** The restriction fragments are separated by size; the distance migrated by a fragment during electrophoresis is inversely proportional to its size.

(d) **DNA denaturation:** The DNA fragments in the gel are made single-stranded.

(e) **DNA transfer (blotting):** DNA is transferred from the gel to the surface of a membrane, such as nitrocellulose. The method of transfer shown here is called capillary blotting.

(f) **DNA immobilization:** The membrane is baked to irreversibly bind the DNA to the membrane.

(g) **Hybridization:** The membrane is submerged in a solution containing many molecules of a specific single-stranded DNA "probe," labeled in some way for later detection. The probe DNA forms base pairs with target DNA molecules on the membrane.

(h) **Washing:** Probe that is not extensively base-paired to the immobilized DNA is washed away; probe that is nonspecifically bound is removed.

(i) **Development/detection:** Restriction fragments that have hybridized with probe appear as a pattern on the membrane (or on the film if the label was a radioisotope).

Figure 31.3 (opposite page) Overview of Southern blotting and hybridization. With the completion of the Southern technique, what was once visible only as a smear of DNA fragments on a gel now becomes a distinct pattern of specific restriction fragments on a membrane.

Materials

First Session: Bacterial DNA Isolation and Restriction Digestion

Cultures
  E. coli B and *S. marcescens*, each grown
    overnight in 2 ml LB broth and
    then inoculated into 50 ml fresh LB
    for log growth

Media
  LB broth: 10 g bacto-tryptone, 5 g yeast
    extract, 10 g NaCl per liter

Reagents
  TNE (10 mM Tris, pH 8.0, 10 mM NaCl,
    0.1 mM EDTA), autoclaved
  TE (10 mM Tris, pH 8.0, 0.1 mM EDTA),
    autoclaved
  HTE (50 mM Tris, pH 8.0, 20 mM EDTA),
    autoclaved
  2% sarcosyl (N-lauroyl sarcosine) in HTE
  RNase on ice (pancreatic RNase A, 10 mg/ml,
    in TE, preheated to 80°C for 10 minutes to
    inactivate DNases)
  Pronase on ice (10 mg/ml, in TNE, preheated
    to 37°C for 15 minutes to inactivate DNases)

Phenol, equilibrated with 0.5 mM Tris, pH 8.0
Chloroform (chloroform:isoamyl
  alcohol, 24:1)
3.0 M sodium acetate
Isopropanol
70% ethanol
Distilled water, autoclaved
Restriction enzyme and control reaction mixes
  (table 31.1)

Equipment
  37°C bacterial incubator with shaker platform
  Microwave oven
  Water bath or heat block at 37°C
  Water bath or heat block at 50°C

Miscellaneous supplies
  Laboratory marker
  Latex gloves (when handling DNA; to protect
    DNA from deoxyribonucleases on hands)
  Ice
  Microfuge tubes
  Pasteur pipettes/bulb
  1.0 ml serological pipette/pipettor
  Micropipettors/tips (1–10 μl, 10–100 μl,
    100–1,000 μl)

Table 31.1 Components of the Restriction Enzyme Mix and the Control Mix. Add 10 μl of each mix to the corresponding reaction and control tubes. Store mixes on ice.

		EcoRI	No enzyme control
		Use 10 μl restriction mix.	Use 10 μl no enzyme control mix.
Restriction mix components	10× restriction buffer	3 μl	3 μl
	Sterile distilled water	6 μl	7 μl
	EcoRI (10–20 units/μl)	<u>1 μl</u>	<u>0 μl</u>
	Total mix volume	**<u>10 μl</u>**	**<u>10 μl</u>**
	Total reaction volume with 20 μl bacterial DNA	30 μl	30 μl

Second Session: Agarose Gel Electrophoresis, Staining, and Southern Transfer

Reagents
  0.8 % agarose gel prepared with TBE: Tris-Borate-EDTA (108 g Tris-base, 55 g boric acid, 40 ml 0.5 M EDTA, pH 8.0, per liter)
  DNA standard, lambda-HindIII, 1 μg per 30 μl TBE; one per gel
  DNA sample loading buffer (tracking dyes): 0.25% bromphenol blue, 0.25% xylene cyanol, 30% glycerol in distilled water
  DNA Blue InstaStain™
  Denaturing solution (0.5 N NaOH, 1.5 M NaCl)
  Neutralization solution (0.5 M Tris, pH 7.5, 1.5 M NaCl)
  20× SSC (3 M NaCl, 0.3 M sodium citrate), diluted to 10× SSC

Equipment
  Horizontal gel electrophoresis system and power source
  Kitchen sponge (one per gel, for Southern transfer)

Miscellaneous supplies
  Micropipettors/tips (1–10 μl, 10–100 μl)
  125 ml Erlenmeyer flask
  Laboratory marker
  Latex gloves (when handling DNA samples)
  1.0 ml microcentrifuge tubes
  Weigh boat or shallow dish (for staining)
  Optitran BA-S supported nitrocellulose membranes
  3MM chromatography paper

Third Session: Probe Preparation and Southern Hybridization

Reagents
  DIG-High Prime DNA Labeling and Detection Starter Kit I (table 31.2)
  Probe DNA: pBLU digested with HindIII (1 μg in 16 μl distilled, autoclaved water). One probe for every 2 membranes.
  20× SSC (3 M NaCl, 0.3 M sodium citrate), diluted to 2× SSC

Equipment
  Oven set at 80°C
  Oven set at 42°C with a rocker platform covered with bench-coat absorbent paper
  Water bath set at 42°C
  Boiling water bath or heat block set at 100°C

Miscellaneous supplies
  Micropipettors/tips (1–10 μl, 10–100 μl)
  50 ml conical tubes

Fourth Session: Washing and Blot Development

Reagents
  20× SSC (3 M NaCl, 0.3 M sodium citrate), diluted to 2× SSC
  2× SSC, 0.1% SDS
  0.5× SSC, 0.1% SDS

Equipment
  Oven set at 42°C with a rocker platform covered with bench-coat absorbent paper
  Water bath set at 42°C
  Water bath or oven at 68°C
  Bench top rocker or shaker platform

Miscellaneous supplies
  3MM chromatography paper
  Large weigh dishes

Procedure

First Session: Bacterial DNA Isolation and Restriction Digestion

Yesterday, each *E. coli* strain was inoculated into 2 ml of LB for overnight growth at 37°C with shaking. Earlier today, each 2 ml culture was transferred into 50 ml of fresh broth in 125 ml flasks and incubated at 37°C with shaking.

1. Remove a flask of bacteria from the 37°C incubator (the culture is expected to be in the log phase of growth), and pipette 1 ml of it into a microfuge tube. Centrifuge the sample in a microfuge at full speed (14,000 RPM) for 15 seconds. Decant the supernatant into a waste receptacle, and let the liquid drain off onto a tissue. Dispose of the tissue in a biohazard bag.

2. Resuspend the cell pellet in 0.3 ml HTE, mixing until there are no remaining cell clumps.

3. Add 0.35 ml 2% sarcosyl in HTE. Mix well by capping and inverting the tube. Note that the liquid is quite cloudy. Once lysis is complete (after step 4), the liquid will be less cloudy.

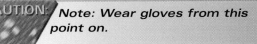
CAUTION: *Note: Wear gloves from this point on.*

Table 31.2 DIG-High Prime DNA Detection Starter Kit I Reagent Descriptions and Buffer Preparations

Reagent	Purpose	Preparation	Amount required (approximate)
Hybridization solution	For prehybridization and hybridization	Add 64 ml of autoclaved, cooled dH_2O, in two portions, stirring, at 37°C for 5 minutes	20 ml per blot
Posthybridization: blot treatment and development			
Buffer 1 (maleic acid buffer)	For preparation of wash buffer and buffer 2 (blocking buffer)	0.1 M maleic acid 0.15 M NaCl pH to 7.5 with solid NaOH	2 liters
Buffer 1 + Tween-20 (wash buffer)	For washing the blot before blocking and after antibody incubation	0.3% Tween-20 (v/v) in buffer 1 Tween-20 (polyoxyethlenesorbitan mono laureate, Sigma # P 1379)	1.5 liters
Buffer 2 (blocking solution)	Coats the membrane with proteins to prevent antibodies from binding directly to the membrane in the fourth session, step 6; also used to make the antibody solution	Dilute 10× blocking solution (provided in kit) 1:10 in buffer 1	75 ml per blot
Buffer 3 (detection buffer)	For equlibration of the blot prior to development, and for preparation of substrate solution	100 mM TrisCl 100 mM NaCl 50 mM $MgCl_2$ ph 9.5	40 ml per blot
Antibody solution	Antibodies, covalently linked to the enzyme AP and specific for the digoxigenin groups along the probe DNA	Anti-DIG-AP (provided in kit) diluted 1:5,000 in buffer 2	10 ml per blot
Substrate solution (blocking solution)	Colorless substrate will be converted to colored product in the presence of AP	200 µl NBT/BCIP (provided in kit) in 10 ml buffer 3	10 ml per blot
Buffer 4 (TE)	For stopping the development reaction	10 mM TrisCl 1 mM EDTA, pH 8.0	20 ml per blot

4. Add 5 µl RNase, and incubate at 37°C for 15 minutes. Add 35 µl of pronase, and heat at 50°C until lysis is complete, about 30 minutes.

5. Cap the tube securely, and vortex the sample for 2 minutes at the highest setting (figure 31.4).

6. *Phenol and chloroform extractions:* Add an equal volume (700 µl) of phenol, shake well, and centrifuge at full speed for 3 minutes to separate the phases. Pipette the upper phase into a fresh microfuge tube, being careful to avoid the

Figure 31.4 Vortex the sample for 2 minutes at the highest setting. This mechanically shears the DNA, generating fragments that are about 20 kilobases (kb). Later, you will further fragment the DNA with the restriction enzyme EcoRI.

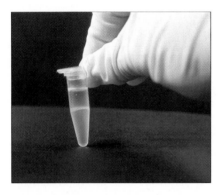

Figure 31.5 Pipette the upper phase into a fresh microfuge tube, being careful to avoid the interface. The interface contains amphipathic substances such as proteins associating with both the aqueous phase above and the organic phase below. The DNA is dissolved in the upper, aqueous phase.

flocculent interface (figure 31.5). Dispose of the phenol waste in an approved receptacle. Extract the sample again with an equal volume of chloroform, centrifuging briefly to separate the phases. Always retain the upper phase and avoid the interface.

7. *DNA precipitation:* Pipette 70 μl of 3M sodium acetate into the sample and mix well. To the mix sample, add an equal volume of isopropanol (700 μl). Mix well by shaking.

8. Centrifuge for 5 minutes at full speed. Look for the pellet as you remove the tube from the centrifuge (figure 31.6). Even if your pellet is not

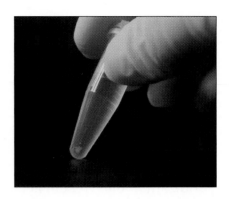

Figure 31.6 The pellet of DNA should be visible as a tiny white clump. Even if you do not see a pellet, the DNA is likely present at the bottom back wall of the tube.

visible at this point, DNA is likely present. Remove as much of the liquid as you can with a Pasteur pipette, being careful not to disturb the DNA pellet. If you do not see a pellet, avoid the back bottom wall of the tube as you pipette.

9. Wash the DNA pellet by adding about 1 ml of 70% ethanol to the tube. Then remove the ethanol without disturbing the pellet. If the pellet comes loose, centrifuge it as in step 8.

10. After removing as much liquid as possible, allow the pellet to air-dry. The pellet will be difficult to see once it is dry, but it is there!

11. Suspend the pellet in 50 μl autoclaved distilled water. Label the tube with your name, the date, and the name of the bacterial strain you used. Store the samples in the freezer, or proceed to the next step.

12. Label two microfuge tubes with your initials. Then label one tube "EcoRI." EcoRI is the name of the enzyme you will be using to digest the DNA. Label the other tube "control."

13. Transfer 20 μl of your DNA sample into each tube. Add 10 μl of restriction mix to the tube labeled "EcoRI" and 10 μl of the no-enzyme control mix to the tube labeled "control" (table 31.1).

14. Mix each sample well by gently pipetting up and down, and place both tubes in a 37°C heat block or water bath for one hour. Samples can also be left overnight at 37°C.

15. Store the samples in the freezer until it is time for the electrophoresis step.

Second Session: Agarose Gel Electrophoresis, Staining, and Southern Transfer

1. Working with one or two other groups, prepare one gel. Weigh out 0.4 g of agarose, and place it into a 125 ml Erlenmeyer flask. Add 50 ml of TBE to the flask, and swirl it gently. Using a lab marker, draw a line on the side of the flask indicating the level of fluid.

2. Microwave the mixture for about 1 minute, checking to make sure it does not boil over. Using a hot glove, gently swirl the flask, and return it to the microwave. Heat for 15 seconds, repeating this until no more flecks of agarose are visible in the flask. If there has been obvious loss of volume through evaporation, add hot distilled water to the flask using the line you drew as a marker. Let the molten agarose cool until the flask is comfortable to handle, but still quite warm.

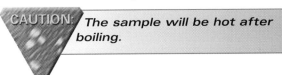 **CAUTION:** *The sample will be hot after boiling.*

3. While you are waiting for the molten agarose to cool slightly, prepare the horizontal electrophoresis chamber according to the manufacturer's instructions. An example of a horizontal minigel system is shown in figure 31.7.

4. When the agarose has cooled as described in step 2, pour the molten agarose, and position the comb. With the long side of the electrophoresis chamber parallel to the edge of the lab bench, the comb should be positioned far to the left. It is important to keep in mind that the samples will run from the black lead end (the negatively charged cathode) toward the red lead end (the positively charged anode).

Figure 31.7 Assembly of a horizontal minigel system (VWR #CBMGU-202). (a) Place dams securely. (b) With the electrode connections toward the back, place the comb so that the comb bar touches the left side dam. Be sure that the teeth of the comb are about 2 mm above the floor of the gel platform, and that the comb is level. (c) When the flask is cool enough to handle, pour the gel.

(a)

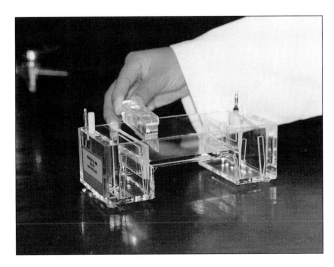

(b)

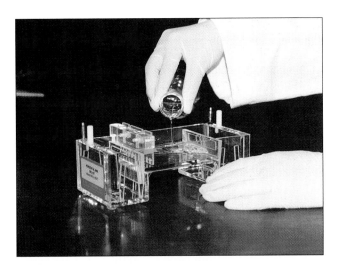

(c)

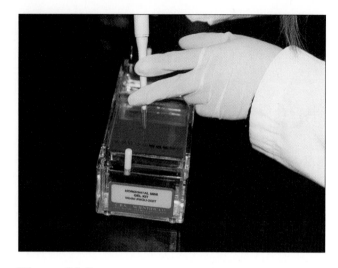

Figure 31.8 Load the agarose gel throught the TBE running buffer. Insert the micropipette tip just inside the well, and gently release the sample. Do not release your thumb from the pipette plunger until you have lifted the micropipettor out of the running buffer.

5. While the agarose is solidifying, prepare the "EcoRI" and "control" samples for loading by adding 6 µl of DNA sample loading buffer. In addition, obtain a DNA standard sample (one per gel) such as lambda-HindIII. Add 6 µl of sample loading buffer to it.

6. When the gel is solid, gently remove the comb and the dams, and pour about 250 ml of TBE into the electrophoresis chamber until the gel is fully submerged.

7. Set a micropipettor at 35 µl. Pipette 35 µl of each sample into its designated well as shown in figure 31. 8, changing the micropipette tip between samples.

8. Place the lid on the electrophoresis chamber, and connect the leads to the power source. Remember that the DNA will migrate from the black lead end toward the red lead end.

9. Set the power source at 90 volts (constant voltage), and allow the electrophoresis to proceed for 1 hour. As the gel begins to run, you will see that the tracking dye is moving toward the red lead end. The dye front allows you to check the progress of the electrophoresis; it does not stain the DNA.

10. Wearing gloves and using a spatula, gently remove the gel from the electrophoresis chamber. Place the gel into a weigh boat or small dish, and stain the gel using the DNA Blue Instastain method. Place a staining sheet over the gel, firmly running your fingers over the surface several times. Then place a glass or plastic plate on top with an empty beaker as a weight, and let the gel and staining sheet set for 15 minutes (figure 31.9).

11. Remove the staining sheet, and place the gel into a shallow dish. Add distilled water heated to 37°C, changing the warm water every 10 minutes until the bands become visible.

12. Examine the banding patterns, comparing the EcoRI-digested and the uncut samples. Diagram your results in your laboratory report. Store the gel wrapped in plastic wrap in the refrigerator, or proceed to the next step.

13. Cut the gel off above the wells (slice through the wells), and notch the gel at its lower left-hand corner (figure 31.10). Measure and record the dimensions of the gel (length and width).

14. Transfer the gel to a small dish containing denaturing solution. Be sure that the entire gel is submerged. Incubate the gel at room temperature for 15 minutes with occasional agitation.

15. Holding the gel in place with a gloved hand, pour the denaturing solution into a beaker, and pour fresh denaturing solution over the gel, submerging it once again. Incubate the gel at room temperature for 15 minutes with occasional agitation.

(a)

(b)

Figure 31.9 Stain the agarose gel after electrophoresis with a methylene blue staining sheet. Make sure there is even contact between the gel and the sheet by (a) running your fingers over the surface several times and (b) placing a plate on top with an empty beaker as a weight.

Figure 31.10 Preparation of the gel for capillary transfer. Cut the gel off above the wells (slice through the wells), and notch the gel at its lower left-hand corner. Then measure the length and width of the gel.

16. Holding the gel in place with a gloved hand, pour the denaturing solution into a beaker, and rinse the gel briefly with distilled water (collect it from a carboy). Holding the gel in place with a gloved hand, pour the distilled water into the sink.

17. Pour neutralization solution into the dish. Be sure that the entire gel is submerged. Incubate the gel at room temperature for 15 minutes with occasional agitation.

18. Holding the gel in place with a gloved hand, pour the neutralization solution into the sink, and pour fresh neutralization solution over the gel, submerging it. Incubate the gel at room temperature for 15 minutes with occasional agitation.

19. During the incubation steps, 14–18, prepare materials for transfer:

    a. Wearing clean gloves, cut a piece of nitrocellulose the *same size* as the gel. Use a razor blade on a cardboard surface. Keep your cut membrane on a clean surface. Notch the membrane at the same position that you notched the gel (lower left-hand corner). Write your initials and the date on the bottom edge with a ballpoint pen.

    CAUTION: *Note: The nitrocellulose membrane should be handled with clean gloves throughout the Southern procedure.*

    b. Using scissors, cut two pieces of Whatman 3MM chromatography paper that are the same size as the gel, and two pieces of paper that are *1 cm larger than the gel* in each dimension.

    c. Cut several paper towels the *same size* as the gel (a 2-inch stack when compressed).

20. Wet the nitrocellulose membrane by flotation in a small dish containing 10× SSC. Once it is wet, submerge it.

21. When the gel has been neutralized (after step 18), set up the transfer as shown in figure 31.11. Allow capillary transfer to proceed overnight.

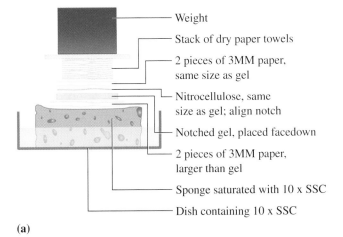

Weight
Stack of dry paper towels
2 pieces of 3MM paper, same size as gel
Nitrocellulose, same size as gel; align notch
Notched gel, placed facedown
2 pieces of 3MM paper, larger than gel
Sponge saturated with 10 x SSC
Dish containing 10 x SSC

(a)

(b)

Figure 31.11 Southern transfer by capillary blotting. (a) Diagram and (b) photograph of the transfer apparatus. The denatured DNA will migrate from the gel onto the nitrocellulose membrane as the salt solution is taken up by capillary action.

Preparation for the Third Session: Disassembly of the Capillary Transfer Apparatus and Membrane Baking

1. Disassemble the transfer apparatus: Throw away the wet paper and gel, and air-dry the nitrocellulose membrane by leaving it on a clean piece of Whatman paper for about 20 minutes.

2. Bake the membrane at 80°C for 1 hour, sandwiched between two pieces of clean Whatman paper with a glass weight on top. Store the membrane, now called the *blot*, the same way at room temperature.

Third Session: Probe Preparation and Southern Hybridization

A summary of the following steps is presented in table 31.3.

1. *DNA probe labeling*

   a. The probe is pBLU cut with HindIII. *Note: Make one probe for every two blots*. Obtain 1 µg of HindIII-cut pBLU DNA suspended in 16 µl of dH₂O. Boil this sample for 10 minutes (or use heat block at 100°C) to denature the DNA. The pBLU DNA molecules must be denatured so they are free to anneal to the random primers and to act as a template for DNA synthesis.

   b. After the 10-minute denaturation step, give the tube a quick spin, and immediately place it on ice.

   c. Add 4 µl of DIG-High Prime (labeling mix) to the denatured DNA, and mix well by gently pipetting up and down. Incubate the sample 1 hour at 37°C.

2. *Prehybridization*

   a. While the probe labeling reaction is going on, wet the nitrocellulose membrane containing DNA by floating it on 2× SSC. Once it is completely wet, submerge it in the 2× SSC.

   b. Transfer 10 ml of hybridization solution (table 31.2) into a 50 ml conical tube, and place into a 42°C water bath.

Table 31.3 The Steps in Southern Hybridization and Development (in Brief)

Step	Description
DNA probe labeling	Single-stranded (denatured) DNA is used as template for the synthesis of labeled DNA. The primers for synthesis are random hexanucleotides, expected to anneal at random sites along the DNA
	During synthesis, dGTP, dATP, dTTP, and dCTP are incorporated along with Digoxigenin-dUTP (the label).
	The probe DNA must be denatured by boiling prior to hybridization.
Prehybridization	The membrane with denatured DNA bound to it is submerged in hybridization solution without the labeled probe. This step helps block the membrane to prevent nonspecific binding of the DNA probe directly to the membrane.
Hybridization	The digoxigenin-labeled DNA probe is added to the membrane in hybridization solution. During hybridization, which typically proceeds overnight, the single-stranded DNA probe binds with complementary sequences of DNA bound to the membrane.
Washing	Washing the membrane removes nonspecifically bound probe.
Antibody incubation	Antibodies specific for the digoxigenin group bind to digoxigenins along the DNA probe. The antibodies are covalently linked to an enzyme, alkaline phosphatase (AP).
Development	The membrane, now containing labeled probe hybridized at specific sites, is placed into a colorless substrate, BCIP/NBT, which is converted to a colored product by the enzyme AP. Color appears only at sites where AP-antibody is located, and the AP-antibody is located wherever digoxigenin (probe) is hybridized.

Figure 31.12 Place the blot into a 50 ml conical tube containing warm DIG-Easy hybridization solution with the DNA side toward the center of the tube.

c. Place the blot into the 50 ml conical tube containing warm DIG-Easy hybridization solution with the DNA side toward the center of the tube (figure 31.12).

d. Place the securely capped conical tube on a rocker platform covered with bench-coat absorbent paper. Incubate with rocking at 42°C until the probe is ready (30 minutes).

3. *Hybridization*

a. Heat the labeled pBLU probe for 10 minutes in a boiling water bath.

b. Give the tube a quick spin, and add 10 µl to the 50 ml conical tube containing your blot and hybridization buffer. Return the conical tube to the oven, and incubate at 42°C with rocking until the next session.

Fourth Session: Washing and Blot Development

Steps 1–3 are designed to remove nonspecifically bound probe. Steps 4–10 are designed for membrane development

1. Remove the hybridized blot from the oven, and turn the oven temperature up to 68°C.

2. Decant the hybridization solution into a waste receptacle, and wash the blot by adding 40 ml of 2× SSC, 0.1% SDS wash to the conical tube. Keeping the tube at room temperature, mix it occasionally over the course of 5 minutes. Decant the solution, and repeat the wash with fresh 2× SSC, 0.1% SDS.

3. Decant the 2× SSC, 0.1% SDS wash, and add 40 ml of warmed 0.5× SSC, 0.1% SDS to the conical tube. Return the tube to the oven, now at 68°C, for 15 minutes with rocking. Decant the solution, and repeat the wash with fresh 0.5× SSC, 0.1% SDS.

4. Place the blot into a weigh dish with the DNA side up. Wash the membrane with 20 ml of buffer 1 containing 0.3% Tween-20 for 1 minute at room temperature with rocking.

5. Holding the blot in place with a gloved hand, decant buffer 1/Tween-20. Transfer 50 ml of buffer 2 into the dish, covering the blot completely. Incubate the blot for 30 minutes at room temperature with rocking.

6. Decant buffer 2, and transfer 20 ml of prepared antibody (alkaline phosphatase-conjugated anti-digoxigenin antibody diluted 1:5,000 in buffer 2 into the dish, covering the blot. Incubate at room temperature for 15 minutes with rocking.

7. Decant the antibody, and wash the blot with 50 ml of buffer 1 + Tween-20 for 15 minutes at room temperature with rocking. Repeat with fresh buffer 1.

8. Decant buffer 1 + Tween-20, and add 20 ml of buffer 3. Gently swirl the dish for 2 minutes.

9. Decant buffer 3 and transfer 10 ml of freshly prepared substrate solution (200 µl NBT/BCIP stock in 10 ml buffer 3). Place the dish in a dark place such as a drawer. No rocking is necessary.

10. Within 3 to 10 minutes, purple-gray bands should appear on the blot. When bands have developed, but before the membrane itself begins to discolor, stop the reaction by adding 50 ml of buffer 4 to the dish. After 5 minutes, decant the solution, and add distilled water to the dish. Pick up the blot, and place it on a clean piece of Whatman paper, allowing it to air-dry. Store the membrane flat, sandwiched between two pieces of Whatman paper, with a weight on top.

11. Record your results in your laboratory report.

LABORATORY REPORT

NAME ——————————————— DATE ——————————

LAB SECTION ——————————————————————————

Bacterial DNA Isolation and Southern Analysis

1. Diagram the banding pattern of your stained gel (or place a photograph of your gel here). Number each lane of the gel. Below the gel diagram or photo, list the lane numbers and what you loaded into each lane.

2. Describe any differences you see in the restriction enzyme–digested sample compared with the control sample.

3. If you completed the Southern portion of the lab, diagram your results in the blank space in question 1, right, and indicate the contents of each lane. Can you distinguish *lacZ*-specific restriction fragments? If so, how many fragments do you see? Do you think that the probe hybridized to other regions of DNA in the genome or to the bacteriophage lambda DNA fragments? If so, this is known as nonspecific hybridization.

4. The rate at which a DNA fragment migrates on a gel during electrophoresis is inversely proportional to the log of its molecular weight. Given this fact, where on the gel are the largest fragments, and where are the smallest fragments?

5. If DNA from a cell is cut with a restriction enzyme and loaded onto a typical agarose gel, only a smear of DNA is seen on a stained gel. How does using the Southern technique overcome this limitation?

6. In a Southern blot, the consequences of not denaturing the DNA in the gel are the same as the consequences of not boiling the probe before adding it to the hybridization solution. Please explain.

Mutagenesis in Bacteria: The Ames Test

Background

An animal or plant cell becomes cancerous when it accumulates mutations that lead to unregulated cell division, chromosomal instability, and/or the inability to undergo normal cell death (*apoptosis*). Therefore, any natural or synthetic agent that damages DNA is a potential carcinogen. In 1971, Dr. Bruce Ames developed a rapid method for identifying mutagens—and so, potential carcinogens—using a special strain of *Salmonella enterica* (formerly *S. typhimurium*). The strain has two features that make it ideal as a sensor for mutagens. First, it lacks DNA repair enzymes so that mistakes in DNA synthesis are not corrected. Second, it carries a point mutation that renders it a histidine auxotroph (*his⁻*); it is unable to synthesize this amino acid from ingredients in its culture medium. In the presence of a mutagen, reversions or *back mutations* to the *his⁺* phenotype occur at a high rate, and the revertants are easily identified.

In the Ames test, the auxotrophic strain is exposed to a test chemical and cultured on a nutrient medium containing only a small amount of histidine. The *his⁻* cells can survive until their histidine is used up. Cells that have reverted to the *his⁺* phenotype continue to grow even in the absence of exogenous histidine. The number of colonies on the test plate is therefore proportional to the efficiency of the mutagen. For example, as shown in figure 32.1, substance A produced a higher frequency of reversion than the control, while substance B did not. The results suggest that substance A is a mutagen but substance B is not.

This bacteria-based mutagenesis test provides a fast, inexpensive way to identify potential carcinogens. It is important to note, however, that some substances that cause cancer in laboratory animals are not muta-

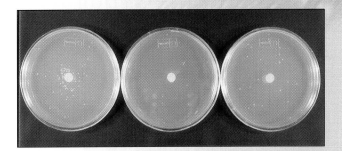

Figure 32.1 An example of Ames test results. The concentration of the amino acid histidine is limiting in each plate, so only *his⁺* revertants grow. The control plate is at the center. Substance A produced a higher frequency of reversion than the control, while substance B did not. The results suggest that substance A is mutagenic and substance B is not.

genic in the Ames test, and some substances identified as mutagens in the Ames test do not appear to cause cancer. Some chemicals (called pro-mutagens) are not mutagenic unless they are converted to more active derivatives by liver enzymes. For example, benzo[a]pyrene is not mutagenic, but it is converted by liver enzymes to diolepoxides, which are potent mutagens and carcinogens. Therefore, to test for pro-mutagens, an extract of rat liver enzymes is usually included in the Ames test.

Since *Salmonella* is pathogenic in humans, we will be using a harmless strain of *E. coli* that is auxotrophic with respect to histidine (and thiamine) as our mutagen-sensor strain. Although this strain is not optimized for mutagenesis (it is capable of DNA repair), the principle of the test is the same. In addition, we will not include liver enzymes in the test, so we will not be testing for pro-mutagens.

Materials

Cultures
   Overnight culture of *E. coli* strain AB 3612
      in nutrient broth
   or *S. typhimurium*, Ames test strain

 All agents in red are BSL2 bacteria

Media
   10 minimal medium agar plates (per group)
      40 plates:
      Sodium phosphate dibasic, 6 g
      Potassium phosphate monobasic, 3 g
      Sodium chloride, 0.5 g
      Ammonium chloride, 1 g
      15 g agar
      1 liter distilled H_2O
      After autoclaving, add 50 ml warmed, sterile
         40% glucose, and swirl gently to mix.
Reagents
   10× thiamine solution (20 mg/ml)
   Sterile distilled water (for water-soluble solids
      to be tested)

Chloroform (for water-insoluble solids
   to be tested)
70% ethanol in a shallow dish
Test substances provided in the
   laboratory (such as diethyl sulfate,
   4-nitro-o-phenylenediamine or sodium
   nitrite) and those supplied by students (such
   as household products). The effects of UV
   radiation can also be tested if a UV lamp is
   available, along with UV-safe goggles
   and gloves.

Equipment
   37°C incubator with shaker platform
   Bunsen burner

Miscellaneous supplies
   Sterile Pasteur pipettes/bulb or
      transfer pipettes
   Microfuge tubes (~8)
   1.0 ml serological pipette/pipettor
   Micropipettors/tips (100–1,000 µl)
   Spreader
   Sterile forceps
   Sterile filter paper disks (0.75 cm diameter)
   Laboratory marker

Procedure

1. Obtain ten minimal medium agar plates: four plates for each substance you are testing (2 substances) and two control plates. Pipette 0.1 ml of 10× thiamine solution onto each of the minimal medium agar plates. Distribute the liquid as evenly as possible with a sterilized spreader (figure 32.2).

2. Once the plates have dried, pipette 0.1 ml of the overnight culture of *E. coli* strain AB 3612 onto each plate with a sterilized spreader. Spread the cells as evenly as you can. Label the plate bottoms with your name(s) and the date.

3. While the plates are drying, prepare two substances that you wish to test for mutagenicity. If the material is a solid, weigh out 1 mg using an analytical balance, place it into a microfuge tube, and dissolve it in 1 ml of sterile, distilled water. *Note:* If the substance does not dissolve in water, weigh out another milligram, and dissolve it in 1 ml chloroform. If the substance is a liquid, record its concentration, if known.

4. *Prepare dilutions of both liquids:* For each substance to be tested, label three microfuge tubes with the name of the test substance, and number them 2, 3, and 4 (tube 1 is the original, undiluted sample). Pipette 1.0 ml of the appropriate diluent (chloroform or sterile water) into the tubes numbered 2, 3, and 4. **If you use chloroform, keep the tubes capped.** Then, using a micropipettor, transfer 1 μl of the undiluted liquid sample into tube 2, and mix well. Using the same tip, transfer 1 μl of sample from tube 2 into tube 3, and mix well. Using the same tip, transfer 1 μl of sample from tube 3 into tube 4, and mix well. Repeat this series of dilutions on the second liquid substance.

5. Label the plates the same way you labeled the microfuge tubes (1–4 and substance name). Label the two remaining plates "dry disk control" and "solvent control."

6. Using sterile forceps, place a sterile filter paper disk at the center of each plate (figure 32.3).

7. Using a sterile Pasteur pipette or transfer pipette, add 1 drop of a liquid sample to the center of the filter paper disk on the corresponding plate. The filter paper should be saturated but not dripping wet. If needed, add additional sample, drop by drop, until the paper is saturated. Count the number of drops you use.

8. To the "dry disk control" plate, add no liquid. To the "solvent control" plate, add either sterile water or chloroform, dropwise, as in step 7. If you used both solvents, choose just one, but be sure that someone else in the class performs the other solvent control.

9. Place the plates into the 37°C incubator, inverted. Be sure that the disk continues to adhere to the agar. Incubate the plates for 2 days (the plates will then be stored in the refrigerator).

10. Examine your plates, and record the results in your laboratory report.

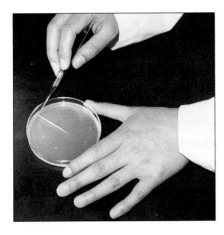

Figure 32.2 Spread 0.1 ml (100μl) 10× thiamine solution onto a minimal medium agar plate.

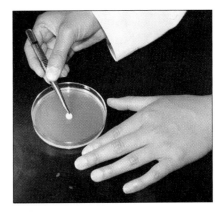

Figure 32.3 Place a sterile filter paper disk at the center of the agar plate.

LABORATORY REPORT

32

NAME _____ DATE _____

LAB SECTION _____

Mutagenesis in Bacteria: The Ames Test

1. Complete the following data tables.

Test substance A:	Substance description:			Concentration (if known) in mg/ml:
Sample number (1–4)	**Sample dilution**	**Total material tested (mg)**	**Description of results (number and distribution of any colonies)**	**Mutagenic at this level?**

Test substance B:	Substance description:			Concentration (if known) in mg/ml:
Sample number (1–4)	**Sample dilution**	**Total material tested (mg)**	**Description of results (number and distribution of any colonies)**	**Mutagenic at this level?**

2. Briefly discuss the results of the Ames test for each of the substances you tested in light of the data you have gathered, comparing these results with the controls.

3. An overnight culture is expected to be at the stationary phase of growth and at a density of about 10^9 cells/ml. Given this, approximately how many cells did you plate initially?

4. If you detected revertant colonies on any of the plates, select one plate, and do the following:

   a. Calculate the approximate surface area (cm^2) of the region where the colonies appear.

   b. What percentage of the total plate surface area does the affected area (determined in 4a) represent?

   c. Given your answer to question 3, calculate the number of cells that you plated in the affected area.

   d. What is the "mutagenesis efficiency" of this substance, expressed as the number of revertant colonies per total number of cells plated?

5. What conclusion would you reach if you observed the following: 20 scattered colonies around the disk of a test plate and 18 scattered colonies around the disk of the "solvent control" plate.

Plasmid Isolation and Restriction Mapping

Background

The capacity of a bacterium to respond to environmental conditions, to reproduce, or to cause disease depends on the expression of its genes. Most of the genes are located on a single, circular, double-stranded DNA (**deoxyribonucleic acid**) molecule—the bacterial **chromosome** (see Exercise 31). However, some bacteria also harbor several copies of a much smaller circular, double-stranded DNA molecule called a **plasmid,** or *episome* (figure 33.1). Plasmids contain genes that are not necessary for day-to-day metabolic processes, but that confer specialized functions, such as the ability to transfer DNA to another bacterium (in the case of a plasmid called a fertility factor, or F factor) or to produce toxins or antibiotic resistance factors.

Plasmids are particularly valuable to a bacterium because they can be present in multiple copies. While a single bacterium has just one chromosome (or two, if it is about to undergo binary fission), it can have as many as 200 copies of a plasmid. Thus, plasmids can offer as many

as 200 copies of a gene encoding an antibiotic resistance factor, for example. Plasmids can be replicated because they have a site for DNA polymerase binding, called the **origin of replication** (ORI or rep). They are replicated more rapidly than the chromosome because they are so much smaller. Figure 33.2 presents a comparison of the features of plasmid and bacterial chromosomal DNA.

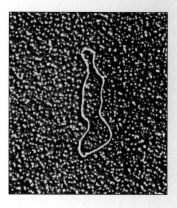

Figure 33.1 Electron micrograph of a plasmid.

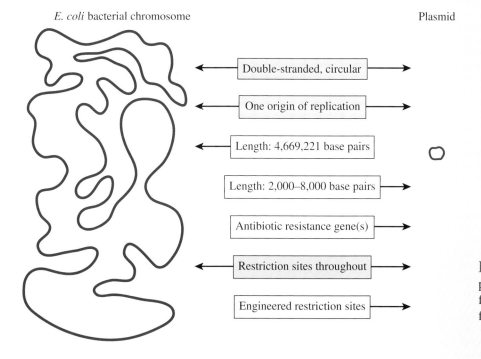

E. coli bacterial chromosome

Plasmid

← Double-stranded, circular →

← One origin of replication →

← Length: 4,669,221 base pairs

Length: 2,000–8,000 base pairs →

Antibiotic resistance gene(s) →

← Restriction sites throughout →

Engineered restriction sites →

Figure 33.2 A comparison of plasmid and bacterial chromosome features. The highlighted boxes are features common to both.

While plasmids occur naturally, they have also been modified for use in gene cloning and gene transfer. Such plasmids have useful marker genes (including antibiotic resistance genes) and *restriction sites* for the insertion of foreign DNA. A particular restriction site is recognized by a particular *restriction endonuclease*, a type of enzyme that occurs naturally in bacteria, which protects bacteria from foreign DNA, typically bacteriophage (virus) DNA. Thus, the term restriction endonuclease makes sense; the enzyme *restricts* the growth of viruses in bacteria by *cutting* double-stranded *DNA* (nuclease) *within* a DNA molecule (endo). There are now hundreds of restriction enzymes available for research. Just a few of them, along with their recognition sites, are listed in table 33.1.

In this exercise, you will isolate the plasmid pBR322 from *E. coli*, cut the plasmid with two different restriction enzymes, electrophorese the *restriction fragments* on an agarose gel, analyze the gel to determine the fragment sizes (lengths in base pairs), and formulate a **restriction map**, showing the relative positions of the restriction sites and the distances between each site. A genetic map of pBR322 is presented in figure 33.3

Table 33.1 Examples of Restriction Endonucleases and Their Recognition Sites*

Restriction enzyme	Bacterial source	Recognition site	DNA ends resulting from restriction
BamHI	*Bacillus amyloliquefaciens* H	↓ —GGATCC— —CCTAGG— ↑	—G GATCC— —CCTAG G—
EcoRI	*Escherichia coli*	↓ —GAATTC— —CTTAAG— ↑	—G AATTC— —CTTAA G—
HindIII	*Haemophilus influenzae* Rd	↓ —AAGCTT— —TTCGAA— ↑	—A AGCTT— —TTCGA A—
NruI	*Nocardia rubrai*	↓ —TCGCGA— —AGCGCT— ↑	—TCG CGA— —AGC GCT—
PstI	*Providencia stuartii*	↓ —CTGCAG— —GACGTC— ↑	—CTGCA G— —G ACGTC—

*Each restriction enzyme was isolated from bacteria, and each recognition site is composed of a molecular palindrome; it reads the same on the upper strand left to right as on the lower strand right to left. The positions at which the enzyme cuts the DNA are indicated by an arrow (↑). Notice that the restriction enzyme cuts at equivalent positions on the two strands.

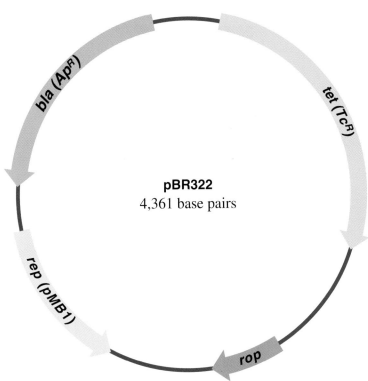

Figure 33.3 A genetic map of the plasmid pBR322 which confers resistance to both tetracycline and ampicillin. The region called *rep* contains the origin of replication The gene *rop* encodes a protein that helps regulate replication.

Materials

First Session: Plasmid Isolation ("Plasmid Miniprep") and Restriction Digestion

Cultures
  E. coli RR1 (wild-type strain, transformed with pBR322), grown on an ampicillin-containing LB agar plate (1 plate per pair)

Media
  LB agar: 10 g bacto-tryptone, 5 g yeast extract, 10 g NaCl, 12 g agar per liter, ampicillin 100 µg/ml

Reagents
  Solution I: 25 mM Tris-Cl, pH 8.0, 50 mM glucose, 10 mM EDTA
  Solution I with lysozyme, 4 mg/ml
  Solution II: 0.2 N NaOH, 1.0% SDS
  Solution III: 3 M potassium acetate (120 ml 5M potassium acetate plus 23 ml glacial acetic acid, bring volume to 200 ml)

Isopropanol
TE/RNase: Tris-EDTA (10 mM Tris, pH 8.0, 1 mM EDTA), 50 µg/ml RNase, heat-treated to denature DNAses
Restriction endonuclease EcoRI and restriction endonuclease PstI, prepared as restriction mixes E, P, and E+P (table 33.2)

Equipment
  Microcentrifuge
  Vortexer
  37°C heat block or water bath

Miscellaneous supplies
  Laboratory marker
  Latex gloves (to protect DNA from deoxyribonucleases on hands)
  Ice
  1.5 ml microfuge tubes
  Pasteur pipettes/bulb
  1.0 ml serological pipette/pipettor
  Micropipettors/tips (1–10 µl, 10–100 µl, 100–1,000 µl)
  Plastic ruler

Second Session: Agarose Gel Electrophoresis

Reagents
Agarose
TBE:Tris-Borate-EDTA (108 g Tris-base, 55 g
boric acid, 40 ml 0.5 M EDTA, pH 8.0,
bring volume to 1 liter)
DNA sample loading buffer (tracking dyes):
0.25% bromphenol blue, 0.25% xylene
cyanol, 30% glycerol in distilled water
DNA standard, lambda-HindIII, 1 μg per 25 μl
TBE; one per gel
DNA Blue InstaStain™

Equipment
Microwave oven
Horizontal gel electrophoresis system and
power source

Miscellaneous supplies
Latex gloves
Micropipettors/tips (1–10 μl, 10–100 μl)
125 ml Erlenmeyer flask
Bacterial waste beaker
Semilog graph paper

Procedure

First Session: Plasmid Isolation and Restriction Digestion

1. Pipette 200 μl of solution I into a 1.5 ml
   microfuge tube. Scrape a large loopful of *E. coli*
   RR1 (wild-type) cells from the plate, and mix
   into the solution, tapping the loop in order to
   release the cells. Mix the sample until there are
   no clumps by pipetting up and down or by
   vortexing the capped tube.

2. Centrifuge the sample for 15 seconds in a
   microcentrifuge to pellet cells.

3. Pipette off the liquid supernatant above the pellet.
   Remove as much of the liquid as you can, and
   discard it into a bacterial waste beaker.

4. Add 200 μl of solution I containing lysozyme (4
   mg/ml) to the cell pellet. Pipette up and down to
   resuspend the cells well.

5. Place the sample on ice for 1 minute.

6. Add 400 μl of solution II. Mix gently by inverting
   several times, and place on ice for 1 minute.

7. Add 300 μl of solution III (cold). Cap the tube
   securely, and vortex the sample for 10 seconds at
   the highest setting. You will see a white

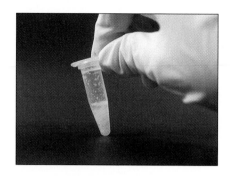

Figure 33.4 A flocculent precipitate forms after the addition of solution III and vortexing.

precipitate forming. Place the sample on ice for 5
minutes. This is bacterial chromosomal DNA,
RNA, proteins, and bacterial debris (figure 33.4).

8. Centrifuge the sample at 14,000 RPM for 5
   minutes at 4°C.

9. Using a Pasteur pipette, transfer the supernatant
   (which contains the plasmid) into a fresh 1.5 ml
   centrifuge tube, avoiding bacterial debris. If you
   transfer debris, centrifuge the sample a second
   time in the fresh tube. In the end, transfer the
   precipitate-free supernatant into a fresh tube.

10. Add 700 μl of isopropanol (a volume equal to the
    sample volume). Shake vigorously. After about 1
    minute, centrifuge the sample for 5 minutes.

11. Look for a tiny white pellet as you remove the
    tube from the centrifuge. It will be located on the
    back side of the tube bottom. Using a Pasteur
    pipette, remove and discard the supernatant,
    avoiding the plasmid pellet.

12. Wash the pellet with a little squirt of 70% ethanol,
    then pipette the ethanol back off and discard it,
    being careful not to discard the pellet. If the pellet
    begins to float, centrifuge the sample again.

13. Remove as much liquid (ethanol) as possible, and
    air-dry the pellet.

14. Suspend the DNA pellet in 50 μl TE containing
    RNAse (50 μg/ml).

15. Label tube with your name, the date, and the
    contents (pBR322).

16. Pipette 7 μl of plasmid DNA into each of three
    fresh microfuge tubes, and place them on ice.
    Label the tubes E (for EcoRI), P (for PstI) and
    E+P (for both). Write your initials on the tubes as
    well. The remaining DNA can be stored in the
    refrigerator or freezer.

17. Add 23 μl of restriction enzyme mix to each
    plasmid sample as shown in table 33.2.

Table 33.2 Components of the Restriction Mix. Add 23 μl of each restriction mix to the corresponding reaction tube.

		Tube E	Tube P	Tube E + P
		23 μl E restriction mix	23 μl P restriction mix	+23 μl E + P restriction mix
Restriction mix components	10× restriction buffer	3 μl	3 μl	3 μl
	Sterile distilled water	19 μl	19 μl	18 μl
	EcoRI	1 μl	0	1 μl
	PstI	0	1 μl	1 μl
	Total restriction mix	23 μl	23 μl	23 μl
	Total reaction volume with 7 μl plasmid	30 μl	30 μl	30 μl

18. Mix the sample well by pipetting up and down gently a few times. If needed, centrifuge for a moment to bring the liquid to the bottom of the tube.

19. Incubate the samples at 37°C for at least 1 hour. They can be left longer, but should not be left overnight. After incubation, store the digested DNA in the refrigerator or freezer.

Second Session: Agarose Gel Electrophoresis

1. Weigh out 0.4 g of agarose, and place it into a 125 ml Erlenmeyer flask. Add 50 ml of TBE to the flask, and swirl it gently. Using a lab marker, draw a line on the side of the flask indicating the level of fluid. Microwave it about 1 minute, checking to make sure it does not boil over. Return the flask to the microwave, and heat again as needed until there are no more flecks of agarose in the flask. If there has been obvious loss of volume through evaporation, add hot distilled water to the flask using the line you drew as a marker. Let the molten agarose cool until the flask is comfortable to handle, but still quite warm.

2. While the agarose is cooling, prepare the horizontal electrophoresis chamber according to the manufacturer's instructions (see figure 31.7).

3. When the molten agarose has cooled slightly, pour the gel and position the comb. With the long side of the electrophoresis chamber parallel to the edge of the lab bench, the comb should be positioned far to the left. It is important to keep in mind that the samples will run from the black

lead end (the negatively charged cathode) toward the red lead end (the positively charged anode).

4. The agarose will solidify as it cools, within about 15 minutes. While the gel is solidifying, prepare your samples for loading. You have three digests, labeled E, P, and E + P. To each of these tubes, add 6 μl of sample loading buffer. In addition, prepare a sample of undigested pBR322 by pipetting 7 μl into a fresh microfuge tube, adding 23 μl TBE and 6 μl of sample loading buffer.

5. Each gel must also contain a size marker, or DNA standard. The standard (here, lambda [λ] DNA cut with HindIII) is a set of fragments of known lengths (figure 33.5). Later you will use the standard to deduce the lengths of your restriction fragments. Add 6 μl of sample loading buffer to a 24 μl sample of standard.

6. When the gel is solid, gently remove the comb and the dams, and pour about 250 ml of TBE into the electrophoresis chamber until the gel is fully submerged.

7. Set a micropipettor at 35 μl. Load 35 μl of each sample into its designated well, changing the micropipette tip between samples. Load in this order:

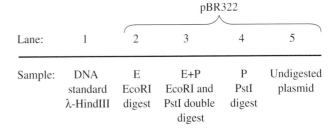

λ-HindIII fragments

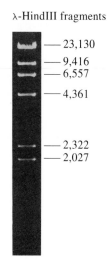

— 23,130
— 9,416
— 6,557
— 4,361

— 2,322
— 2,027

Figure 33.5 Bacteriophage lambda (λ) DNA digested with HindIII yields eight restriction fragments. Because the lengths of the fragments are known, they can be used as a size standard. Here, the two smallest fragments are not visible.

8. Place the lid on the electrophoresis chamber, and connect the leads to the power source. Remember that the DNA will migrate from the black lead end toward the red lead end.

9. Set the power source at 80 volts (constant voltage), and allow the electrophoresis to proceed for about 1 hour. As the gel runs, you will see that the tracking dyes are moving toward the red lead end as well. The dye fronts allow you to check the progress of the electrophoresis. The dye does not indicate the position of DNA fragments.

10. After 1 hour, turn off the power. Wearing gloves and using a spatula, gently remove the gel from the electrophoresis chamber. Place the gel onto a piece of plastic wrap, and stain the gel using the DNA Blue InstaStain method. Place a staining sheet over the gel, firmly running your fingers over the surface several times. Then place a glass or plastic plate on top of the gel with an empty beaker as a weight, and let the gel and staining sheet set for 15 minutes.

11. Remove the staining sheet, and place the gel into a shallow dish. Add distilled water heated to 37°C, changing the warm water every 10 minutes until the bands become visible. Gels can be left to destain overnight.

12. Using a plastic ruler, measure and record the distance migrated (cm) by each of the standard fragments (in the lamba-HindIII lane). Be sure to use the same start point for each measurement, such as the top end of the gel or the bottom of the well. Record each value in your laboratory report. Then measure and record the distances migrated by your restriction digest fragments in each lane: E, E + P, and P.

13. Using a piece of semilog paper, graph the standard. Plot the distance migrated by each standard fragment on the x (linear) axis versus the log of its length (in base pairs) on the y (log) axis. When you use log paper, you do not need to calculate log. Alternatively, you may use a graphing program to plot the data.

14. Draw the best straight line. Do not include the data points from the largest two standard fragments (23,130 and 9,416). An example of a semilog plot is shown in figure 33.6.

15. Using the distances you recorded for each of the restriction fragment bands, determine their lengths using the standard graph. Include this graph in your laboratory report.

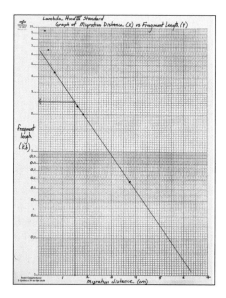

Figure 33.6 Graph of migration distances (cm) versus length in base pairs for the DNA size standard, lambda-HindIII. The red line shows that the length of an unknown fragment can be deduced from its migration distance. Here, a fragment that migrates 1.5 cm is deduced to be approximately 2,500 base pairs long.

LABORATORY REPORT

33

NAME _____ DATE _____

LAB SECTION _____

Plasmid Isolation and Restriction Mapping

1. Complete the following table of DNA standard fragment lengths and migration distances based on your measurements.

Lambdaphage DNA HindIII standard fragment lengths (base pairs)	Migration distance (cm)

2. Using the semilog paper provided, graph the standard fragment lengths versus migration distances as described in the second session, step 13.

3. List the migration distances of the band or bands you measured in each of the pBR322 digest lanes.

EcoRI digest	EcoRI and PstI double-digest	PstI digest

Note: Your gel includes an undigested sample of pBR322 in lane 5. Be sure to compare the bands you see in the digested lanes with those in the undigested lane. If a band in a digested lane matches a strong band in the undigested lane, it may be incompletely digested plasmid and should be ignored.

4. You can make some predictions about the restriction map of pBR322. Draw a circle representing a plasmid.

   a. If you cut a circular DNA at one position, how many fragments will be generated? _____

   b. If you cut a circular DNA at two positions, how many fragments will be generated? _____

5. Consider each of the pBR322 digest results.

   EcoRI

   How many fragments or bands do you see? _____

   Based on this, how many EcoRI sites are there in pBR322? _____

   EcoRI + PstI

   How many fragments or bands do you see? _____

   PstI

   How many fragments or bands do you see? _____

   Based on this, how many PstI sites are there in pBR322? _____

6. Briefly state what you know about the restriction map of pBR322 at this point.

7. Determine the lengths of each of the pBR322 digest fragments using the standard graph you have already prepared. In the space provided, list the fragment lengths for each digest. For each, total the fragment lengths to obtain the total length of the plasmid. Each of the three totals should agree.

   EcoRI fragment(s)   EcoRI + PstI fragment(s)   PstI fragment(s)

   Sum of fragment lengths in each digest (bp)

8. Using the circle you drew in number 4, draw a restriction map of pBR322 providing:

   • the total length of the plasmid (in base pairs)

   • the relative positions of the EcoRI and PstI sites

   • the distance between these sites (in base pairs)

   You can also include the origin of replication and the two antibiotic resistance genes in your map as shown in figure 33.3. Here are some hints on the placement of these sequences: The ampicillin resistance gene is about 1,000 base pairs long, and the PstI site is located within this gene. The tetracycline resistance gene is about 1,200 base pairs long and is located about 300 base pairs to one side of the EcoRI site.

Acquiring Antibiotic Resistance Through Bacterial Transformation

Background

By the 1970s, it appeared that many once-devastating infectious diseases had been all but defeated by antibiotics and highly effective, preventive vaccines. Viral diseases such as poliomyelitis and smallpox were well under control owing to intensive immunization programs, and bacterial diseases such as tuberculosis (TB) were effectively treated with antimicrobial drugs. Since then, these victories, including the eventual eradication of smallpox by 1980, have been greatly tempered by the recent, sharp rise in the rate of infectious diseases worldwide.

Many factors have contributed to the emergence and spread of old, new, and more virulent infectious agents, including climate change, environmental degradation, mass movement of displaced people, international travel, poverty, and the lack of public health measures and surveillance. At the same time, of course, microorganisms have adapted and thrived in new hosts and environments. The adaptability of microbes can be seen in the alarming rise in antibiotic-resistant strains of bacteria, a trend that has been fueled by the misuse of antibiotics in recent decades, and the general sense that bacterial infections are treatable and are therefore of little consequence. In fact, we commonly encounter bacteria that are resistant to more than one antibiotic, the so-called multidrug-resistant, or MDR strains.

A bacterium can acquire resistance to an antibiotic by random, spontaneous mutation within its genome, or by taking in whole antibiotic resistance factor–encoding genes from other microbes. Bacteria can take up foreign genes in one of three ways: transduction, conjugation, or transformation. In **transduction**, a piece of bacterial DNA is transported from one bacterium to another by a **bacteriophage**, a virus that infects bacteria. **Conjugation** involves the direct transfer of DNA from one bacterium to another through an appendage called the *sex pilus*. This mode of natural gene transfer is described in more detail in Exercise 35. Both conjugation and transduction can result in changes in the recipient cell because they involve the transfer of genes. You could say that the recipient cell can be "transformed" from one phenotype to another—for example, from being antibiotic-sensitive to being antibiotic-resistant.

The term **transformation**, however, is reserved for the third mode of bacterial gene transfer: the uptake of free DNA from the surrounding environment. In the late 1920s, the English biochemist Frederick Griffith discovered what came to be known as transformation by chance while working to develop a pneumonia vaccine. Griffith found that a nonvirulent form of *Streptococcus pneumoniae* became virulent by taking up material (later shown to be DNA) from dead, virulent streptococci (figure 34.1).

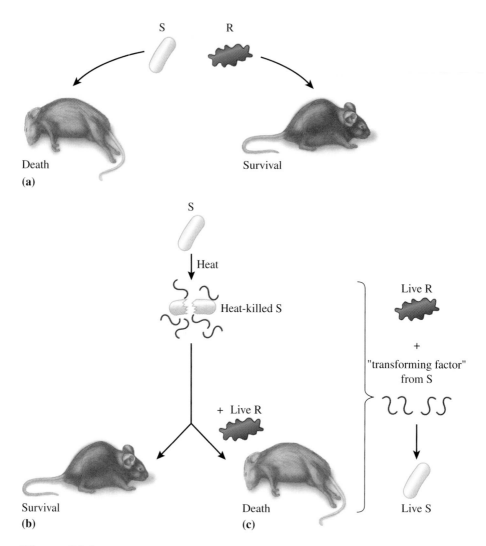

Figure 34.1 The transforming principle. *Streptococcus pneumoniae* exists in two forms: the **S**, or smooth, form is highly virulent because it bears a capsule that resists phagocytosis. On the other hand, the **R**, or rough, form does not cause disease because it has no capsule and is readily eliminated by phagocytic cells. (a) When Griffith inoculated mice with the S strain alone, they succumbed to the infection, while those infected with the R strain alone remained alive and healthy. (b) He then inoculated mice with the heat-killed S strain, and as expected, the dead cells could not establish an infection, and the mice lived. (c) However, when heat-killed S cells were mixed with live, nonpathogenic R cells and introduced together, the mice succumbed to the infection, suggesting that the nonvirulent R cells had been transformed into virulent S cells by taking in genetic material released from the dead S cells. Twenty years later (1944), Avery McCarty, and McLeod extended these studies and provided evidence that this genetic material, or "transforming factor," is DNA.

The DNA that enters the cell can remain as a plasmid, independent of the chromosome, or it may be incorporated into the bacterial chromosome. Not all bacteria can take up free DNA this way. Those that can, such as *Streptococcus pneumoniae*, are said to be naturally **competent**. Other bacteria, including *E. coli*, must be treated to become competent for transformation. *E. coli* can be made competent by first suspending the cells in a solution of calcium chloride. The bacterial cell membrane is permeable to chloride ions, but nonpermeable to calcium ions. As chloride ions enter the cells, so do water molecules, causing the cells to swell slightly and become porous. When the cells are then "heat-shocked" (42°C, 2 minutes), free DNA molecules such as plasmids are swept through the transient pores into the cell (figure 34.2).

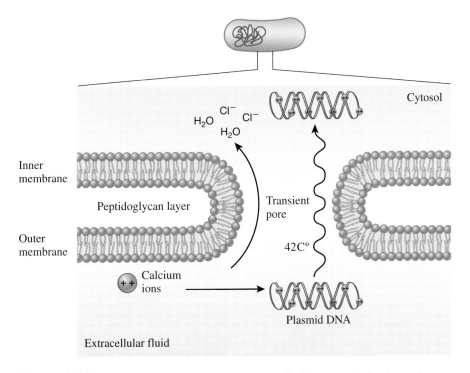

Figure 34.2 The transformation of competent cells. The negatively charged plasmid DNA is associated with calcium ions, while chloride ions and water enter the cell, causing it to swell slightly. During the heat-shock step, the plasmic DNA is swept into the cell. Under the conditions of transformation, typically one plasmid molecule enters a cell.

In this exercise, you will transform antibiotic-sensitive *E. coli* with the plasmid you isolated in Exercise 33, pBR322, yielding transformants resistant to ampicillin and tetracycline (figure 34.3; see figure 33.3 for a map of the plasmid). Here, you will select for transformants by growing the cells on nutrient agar containing ampicillin. The mechanism of ampicillin action as well as the mechanism of resistance to it is shown in figure 34.4.

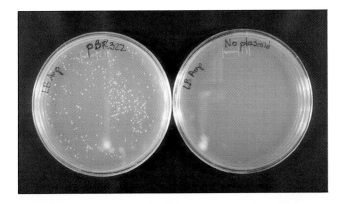

Figure 34.3 pBR322 confers resistance to both ampicillin and tetracycline. *E. coli* transformed with pBR322 grows on an ampicillin-containing plate (*left*). *E. coli* carrying no plasmid does not.

Materials

Cultures
  E. coli RR1 (host strain) mid-log culture
    (10 ml per group)

Media
  LB broth: 10 g bacto-tryptone, 5 g yeast
    extract, 10 g NaCl per liter
  LB agar (12 g/L) plates (100 × 15 mm)
    containing ampicillin at 100 µg/ml
    (2 per group)
  LB agar plates (100 × 15 mm) with no
    antibiotic (2 per group)

Reagents
  50 mM calcium chloride (sterile, cold)
  TE (Tris-EDTA: 10 mM Tris, pH 8.0, 1 mM
    EDTA)
  70% ethanol in a shallow dish

Equipment
  37°C incubator with shaker platform
  Heat block or water bath at 42°C
  Bunsen burner

(a) Peptidoglycan synthesis

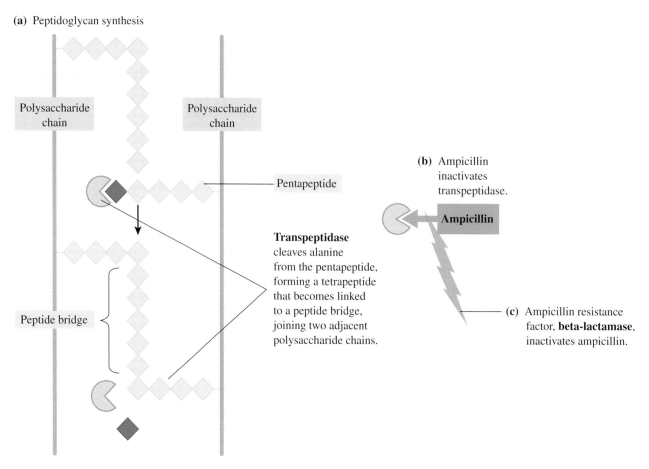

Figure 34.4 The action of ampicillin and ampicillin resistance. (a) Peptidoglycan synthesis. (b) Ampicillin blocks peptidoglycan synthesis. It contains a beta-lactam ring that binds irreversibly to the bacterial enzyme, transpeptidase, blocking a key step in peptodoglycan synthesis. Note that ampicillin does not damage already existing peptidoglycan. (c) The ampicillin resistance factor inactivates ampicillin. It is a beta-lactamase that inactivates ampicillin by breaking its beta-lactam ring.

Miscellaneous supplies
   Ice
   Laboratory marker
   Latex gloves (to protect DNA from
      deoxyribonucleases on hands)
   5 ml pipettes/pipettor
   Micropipettor/tips (10–100 µl)
   1.0 ml serological pipettes/pipettor
   Spreader

Procedure

Prior to today's lab, a 2 ml sample of nutrient broth was inoculated with *E. coli* RR1 (host strain) for overnight growth at 37°C with shaking. Earlier today, 100 ml of nutrient broth was inoculated with the overnight culture and incubated at 37°C with shaking.

Preparation of Competent Cells (yields enough for five transformations)

1. Obtain 10 ml of mid-log *E. coli* RR1 (host strain) cells in a 15 ml conical tube; place it on ice.

2. Pellet the cells by centrifugation at 2,000 RPM ($1,000 \times g$) for 10 minutes at 4°C.

3. Decant the supernatant into a waste receptacle, being careful not to discard the pellet. Leave a small volume of broth over the pellet.

4. Tap the tube vigorously to disperse the pellet in the residual broth. Place the cells on ice immediately.

5. Add 5 ml of sterile, ice cold 50 mM $CaCl_2$ to the cells, resuspending them gently with the pipette.

6. Immediately place the cells on ice for 20 minutes.

7. Centrifuge as in step 2, but for 5 minutes.

8. Decant the supernatant as in step 3. This time, the cell pellet is softer and more diffuse. Be sure it doesn't pour out with the supernatant.

9. Resuspend the cell pellet with 1 ml of ice-cold 50 mM $CaCl_2$, by pipetting the cells up and down very gently. Immediately place these cells back on ice. These are your competent cells.

Transformation of Competent E. coli with pBR322

1. Label two microfuge tubes, "pBR322" and "no plasmid."

2. Place the two labeled tubes on ice, and transfer 200 µl of competent cells into each of the tubes. **Be sure that the cells are well suspended before you transfer them!**

3. Obtain a sample of the plasmid you isolated in Exercise 33. Keep it on ice.

4. Transfer 20 µl of plasmid (~1 µg) into your tube labeled "pBR322." Mix gently, keeping the cells on ice.

5. Using a fresh micropipette tip, transfer 20 µl of TE into your "no plasmid" tube. Mix gently, keeping the cells on ice.

6. Incubate the cells on ice for 20 minutes.

7. While your transformation reactions set on ice, label two ampicillin plates and two antibiotic-free plates with your name and the date. Label one ampicillin plate and one antibiotic-free plate "pBR322." Label the other ampicillin plate and the other antibiotic-free plate "no plasmid control."

8. After the transformation reactions have been on ice for at least 20 minutes, heat-shock each by placing them in a 42°C water bath for 2 minutes. Return them to ice.

9. Add 1 ml of fresh, sterile nutrient broth to each tube, cap tightly, and tape the tubes, side down, onto the shaker platform in the 37°C incubator. Incubate for 40 to 60 minutes with shaking. This allows the cells to recover from the calcium chloride and heat-shock treatment before you plate them. It also allows the antibiotic resistance gene(s) on the plasmid to begin to be expressed, before the cells are exposed to ampicillin.

10. When the recovery period is completed, plate the bacteria by spreading with a sterile spreader:

    100 µl of pBR322 Tf reaction onto the ampicillin plate labeled "pBR322"

    100 µl of pBR322 Tf reaction onto the "no antibiotic" plate labeled "pBR322"

    100 µl of "no plasmid" control reaction onto the ampicillin plate labeled "no plasmid control"

    100 µl of "no plasmid" control reaction onto the "no antibiotic" plate labeled "no plasmid control"

11. Place the inverted plates in the 37°C incubator overnight.

12. Examine the plates and address the questions in your laboratory report.

LABORATORY REPORT

34

NAME _____ DATE _____

LAB SECTION _____

Acquiring Antibiotic Resistance Through Bacterial Transformation

1. Examine each plate. Describe and discuss the results of each with respect to the presence or absence of bacterial growth, and whether or not the plates with growth contain isolated (individual) colonies. Are the results what you expected?

2. Count the total number of colonies on each plate that has individual colonies. *Note:* If the plate is very crowded, it may be easier to count if you divide the plate into quarters or eighths and then multiply the count by 4 or 8, respectively. Record these counts here.

3. Transformation efficiency is a measure of the success of transformation. It is expressed as the number of antibiotic-resistant colonies per μg of DNA transformed. A typical transformation efficiency is about 10^6 colonies/μg. Using the number of colonies on the ampicillin plate labeled "pBR322," determine the transformation efficiency. Keep in mind that you transformed 1 μg of DNA, and that you plated about one-twelfth of the total transformation reaction (100 μl from 1,200 μl total).

4. Do you think the cells that grew on the ampicillin plate are also resistant to tetracycline? Why or why not?

5. Why are most antibiotics safe for humans and other animals (other than side effects) even though they can be very harmful to bacteria?

6. How does ampicillin kill ampicillin-sensitive bacteria? How do ampicillin-resistant bacteria avoid being killed by ampicillin?

7. Why is having an antibiotic resistance gene on a plasmid more beneficial to the bacterium than having the gene on the bacterial chromosome?

8. An isolated colony represents one cell that landed at that spot on the agar when you spread the bacteria. Why is the colony considered a clone?

Acquiring Antibiotic Resistance Through Bacterial Conjugation

Background

Conjugation involves the direct transfer of plasmid or chromosomal DNA from one bacterium to another via an extended appendage called the **sex pilus,** or **conjugation pilus**. The donor cell, also called the male, possesses a plasmid (**fertility factor, or F factor**) that allows the cell to synthesize the sex pilus and to replicate and transport the F factor itself. The recipient cell, or female, is a closely related strain or species (usually Gram-negative) that has a recognition site on its surface.

F⁺ (donor)   F⁻ (recipient)

0 minutes

2 minutes

Pilus formation

10 minutes

DNA replication with
continued pilus formation

15 minutes

DNA transfer

20 minutes

Conjugates separate.

(a)

Specialized conjugation plasmids known as **resistance factors**, or **R factors**, not only carry genes that control conjugation, but can also carry genes that confer resistance to antimicrobial drugs. Thus in a single conjugation event, a recipient cell can receive a "shield" of one or more drug resistance genes. Later, the recipient can spread that resistance to other cells, again through conjugation. Indeed, this type of horizontal gene transfer contributes to the emergence of multidrug-resistant bacteria. Penicillin- and tetracycline-resistant *Neisseria gonorrhoeae* is thought to result from the conjugative transfer of an R factor.

In this exercise, the donor (F⁺) strain is *Escherichia coli* BB4. Its conjugative plasmid carries a tetracycline resistance gene, so the plasmid could be called an R factor. The recipient (F⁻) strain is *E. coli* SCS1. Its chromosome carries the gene for ampicillin resistance. When the cells are mixed and conjugation occurs, a copy of the R factor of *E. coli* BB4 moves to *E. coli* SCS1, conferring tetracycline resistance on the recipient cell (figure 35.1). The success of conjugation can be measured by the appearance of colonies resistant to both ampicillin and tetracycline.

Figure 35.1 Bacterial conjugation. In this example, conjugation is conferred by a fertility factor that carries a tetracycline resistance gene. (a) During conjugation, a copy of the R factor of *E. coli* BB4 moves to *E. coli* SCS1. The SCS1 cell gains tetracycline resistance. (b) An electron micrograph of two *E. coli* cells during conjugation. The F⁺ cell to the right is covered with small pili, and a sex pilus connects the two cells.

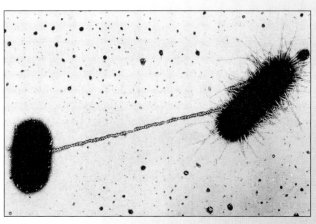

(b)

Materials

Cultures
> Overnight culture of *E. coli* BB4 (tetracycline-resistant donor strain)
> Overnight culture of *E. coli* SCS1 (ampicillin-resistant recipient strain)

Media
> Tryptic soy broth (tryptone 15 g, soytone 5 g, sodium chloride 5 g, in 1 liter distilled water)
>
> Tryptic soy agar plates (tryptic soy broth, agar 15 g/liter)
>> with ampicillin at 100 µg/ml (one per group)
>> with tetracycline at 15 µg/ml (one per group)
>> with ampicillin at 100 µg/ml and
>> tetracycline at 15 µg/ml (2–4 per group)

Reagents
> 70% ethanol in a shallow dish

Equipment
> 37°C incubator with shaker platform
> Bunsen burner

Miscellaneous supplies
> 1.0 ml serological pipettes/pipettors
> Pasteur pipettes/bulb
> 10 ml culture tube (one per group)
> Spreader
> Inoculating loop (for spreading bacteria on a plate half)
> Laboratory marker

Procedure

1. Using a 1.0 ml serological pipette, transfer 0.1 ml of the overnight culture of *E. coli* BB4 (donor cells) into a culture tube. With a fresh pipette, add 0.9 ml of an overnight culture of *E. coli* SCS1 (recipient cells) to the donor cells. Add 5 ml of sterile nutrient broth. Incubate the conjugation mix at 37°C with shaking. Record the time.

2. Obtain one ampicillin plate and one tetracycline plate. Using a lab marker, divide each plate in half by drawing a line on the bottom plate. Label one half of each plate "donor," and the other half of each plate "recipient."

3. Using a sterile Pasteur pipette or a 1.0 ml serological pipette, transfer one drop of the *E. coli* BB4 overnight (be sure that the cells are well suspended first) onto the "donor" half of the ampicillin plate and then onto the donor half of the tetracycline plate. Use a sterile spreader or sterile inoculating loop to spread the bacteria on each plate, being careful not to go beyond your drawn line (figure 35.2).

4. Repeat step 3 using a fresh pipette and transferring a drop of the *E. coli* SCS1 overnight on the "recipient" half of each plate.

5. Near the end of the lab period,* remove the conjugation mix (from step 1) from the incubator, and transfer one drop of the culture onto an ampicillin/tetracycline plate. Use a sterile spreader to distribute the cells evenly over the plate surface. Record the time, and return the conjugation mix to the incubator shaker platform for overnight growth. Label the plate "conjugation mix," and write the time on the plate.

   Note: You may be asked to plate the conjugation mix at earlier time points as well. Label a fresh ampicillin/tetracycline plate for each time point.

   Label all plates with your name(s), the date, and the time of mating period (on the bottom side, along the plate edge). Place all the plate cultures in the 37°C incubator, inverted, overnight.

6. If possible, repeat step 5 the next day.

7. Examine the plate and record your observations in your laboratory report. In addition, count and record the number of colonies on the ampicillin/tetracycline plate(s).

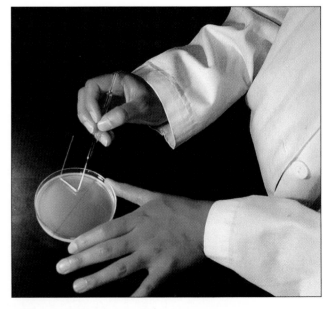

Figure 35.2 Spread the "donor" half of an ampicillin plate with *E. coli* BB4 and the "recipient" half of the plate with *E. coli* SCS1. Do the same with a tetracycline plate.

LABORATORY REPORT

35

NAME _____ DATE _____

LAB SECTION _____

Acquiring Antibiotic Resistance Through Bacterial Conjugation

1. What was the purpose of plating *E. coli* BB4 and *E. coli* SCS1 separately on plates containing ampicillin alone and tetracycline alone? What are the expected results?

2. Diagram and briefly describe the results of the plating addressed in question 1.

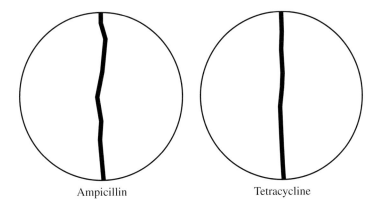

Ampicillin   Tetracycline

3. Formulate a table showing

   • the number of conjugation mixtures you plated

   • the length of the mating period for each

   • the number of colonies you counted on each plate

4. Discuss the results presented in question 3.

5. Approximately what volume of conjugation mixture did you plate at each time point (about how many microliters are there in a drop of fluid)?

    For each time point, what is the conjugation efficiency as expressed in conjugation events per ml? Keep in mind that each colony represents a single cell.

6. Why was it important to use a recipient cell that contained a stable marker such as ampicillin resistance in this experiment?

7. If you were to mix the doubly resistant SCS1 cells with an appropriate, antibiotic-sensitive recipient *E. coli* strain, do you think the recipient strain might become tetracycline resistant? Why or why not? Do you think the recipient strain might become ampicillin resistant? Why or why not?

8. The cells that survive on the ampicillin/tetracycline plates are *E. coli* SCS1, not *E. coli* BB4.
   How do you know this?

9. Two *E. coli* strains, mating pairs X and Z, are mixed. X is F$^+$, harboring an R factor that confers
   resistance to penicillin. Z is F$^-$ and is sensitive to all antibiotics. However, Z has been engineered with a
   gene that makes its colonies appear blue. After 10 minutes, the cells are placed into a blender at high
   speed to disrupt the conjugates.

   One drop of the mating mixture is spread on a plate containing no antibiotic, and another drop is
   spread on a plate containing penicillin. Describe the growth you expect on each plate.
   Briefly explain your answer.

Viruses

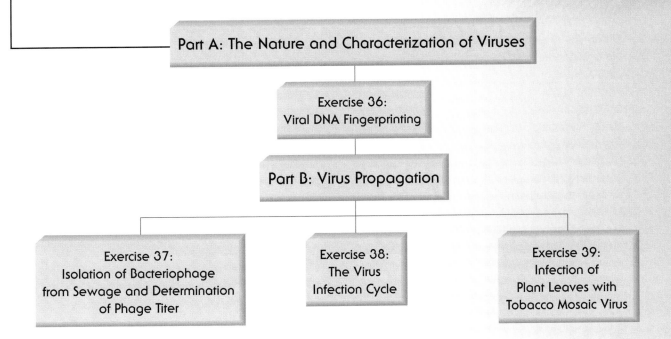

Part A: The Nature and Characterization of Viruses

Exercise 36:
Viral DNA Fingerprinting

Part B: Virus Propagation

Exercise 37:
Isolation of Bacteriophage
from Sewage and Determination
of Phage Titer

Exercise 38:
The Virus
Infection Cycle

Exercise 39:
Infection of
Plant Leaves with
Tobacco Mosaic Virus

36

Viral DNA Fingerprinting

Background

Recall that during the second half of the nineteenth century, Robert Koch, Louis Pasteur, and others began to identify the microorganisms that cause a number of infectious diseases. The researchers relied on the fact that each microbe could be collected in a bacteria-proof filter, grown in nutrient medium, and observed by light microscopy. However, the causes of a number of transmissible diseases, such as foot-and-mouth disease, rabies, and smallpox, remained a mystery. By the dawn of the twentieth century, it had become clear that diseases such as these are associated with a fundamentally different kind of infectious agent—one that is smaller than bacteria (it could not be retained in filters of that time, and it cannot be seen by light microscopy) and is incapable of reproduction outside of cells (cannot be grown in laboratory media).

A **virus** is an obligate intracellular parasite consisting of nucleic acid (an RNA or DNA genome) contained within a coat of proteins called a *capsid* (figure 36.1). More complex viruses contain additional struc-

tures, sometimes including a membranous envelope studded with protein spikes, or *peplomers* (figure 36.2). The power of a virus, an inert, nonliving agent, is in its capacity to enter and be replicated within its host cell. As a result of one virus entering a single cell, hundreds of newly formed viruses may be released as the cell dies. Each virus, or *virion*, is then capable of infecting a nearby cell, effectively extending the cellular injury and leading to symptoms that may arise from the infection or from the immune response to it. In humans, viruses cause a number of diseases, including smallpox (eradicated as of 1980), the common cold, chickenpox, influenza, poliomyelitis, rabies, ebola hemorrhagic fever and AIDS. A few viruses have even been linked to the development of cancer.

In recent years, a number of methods have been developed for the rapid identification of microbes through **DNA fingerprinting,** or **DNA typing**. A DNA fingerprint is a ladder of fragmented or newly synthesized DNA molecules that form a barcode-like pattern unique to an organism. The key to a DNA fingerprint, then, is having an identifiable pattern of bands on a

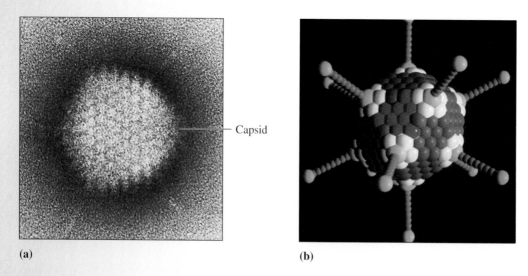

(a)

(b)

Figure 36.1 An adenovirus consists of a DNA genome surrounded by a capsid. Adenoviruses cause upper respiratory infections, such as the common cold, and other infections, such as pinkeye. (a) Electron micrograph of an adenovirus. (b) Computer-simulated model of an adenovirus.

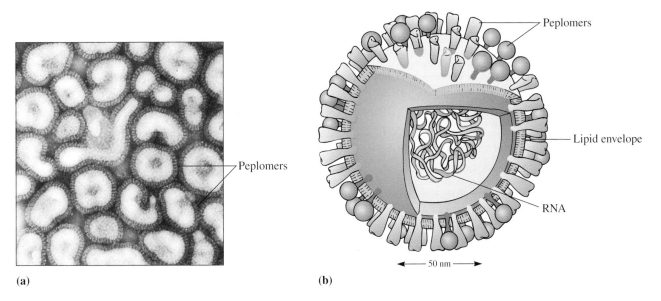

(a)   **(b)**

Figure 36.2 An influenza virus consists of an RNA genome surrounded by a capsid enclosed in a membranous envelope. The envelope contains viral proteins called peplomers, or spikes. (a) Electron micrograph of influenza virus particles. (b) Diagram of an influenza virus.

gel. As you learned in Exercise 31, gel electrophoresis of cellular DNA that has been cut with a restriction endonuclease generates a smear of restriction fragments on a gel. In Exercise 33, however, you saw how the electrophoresis of smaller DNAs, such as plasmids or bacteriophage DNAs, cut with restriction enzymes, can result in a discernible pattern. Among the methods used to generate DNA fingerprints of cellular DNAs are **pulsed-field gel electrophoresis (PFGE)**, the **polymerase chain reaction (PCR)**, and **Southern blotting and hybridization**. In Exercise 31, you learned how a

Southern blot is used to detect particular DNA sequences within a complex genome, such as that of *E. coli*. PFGE and PCR are outlined in figure 36.3.

In this exercise, we will take advantage of the smallness of viral genomes, and generate DNA fingerprints simply through restriction enzyme digestion followed by agarose gel electrophoresis and staining. It will be possible to identify a simulated, unknown "clinical sample" by comparing its restriction pattern with those of known viral DNA samples.

(a) Pulsed-field gel electrophoresis (PFGE)

Isolation of DNA from tissues, cells, or viruses: The DNA is mechanically sheared during this procedure, generating large fragments.

Restriction enzyme digestion: The large fragments of DNA are cut at specific sites with a restriction enzyme, generating restriction fragments characteristic of the organism.

Agarose gel electrophoresis: Very long fragments of DNA (from 40 kb to 5 Mb) are separated by size using alternating electric fields.

Field A on, field B off: DNA migrates on end, parallel to field A, toward the anode (DNA has a net negative charge).

Field B on, field A off: DNA migrates on end, parallel to field B, toward the anode.

Alternate fields several times.

As in standard gel electrophoresis, the distance migrated by a fragment during electrophoresis is inversely proportional to its size. However, PFGE conditions allow for very long fragments to separate by size so they can be distinguished as bands on a stained gel. The restriction fragment lengths, unique to a particular microbe, provide the organism's DNA fingerprint.

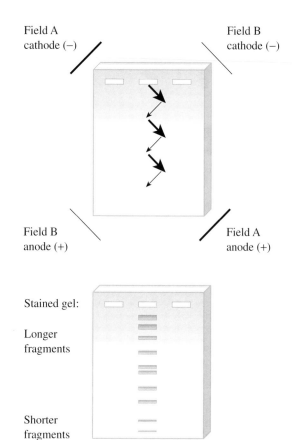

Field A cathode (−)   Field B cathode (−)

Field B anode (+)   Field A anode (+)

Stained gel:

Longer fragments

Shorter fragments

Figure 36.3 Pulsed-field gel electrophoresis and the polymerase chain reaction can be used to generate DNA fingerprints without Southern blotting. (a) PFGE: During typical gel electrophoresis, DNAs longer than 30 kb migrate with the same mobility regardless of size. However, if the DNA is made to change direction during electrophoresis, as in PFGE, these larger fragments can separate from each other. Through PFGE, it is possible to resolve fragments as long as 40,000–5,000,000 base pairs. (b) PCR: Using DNA encoding 16S rRNA as a template, 16S rDNA primers can be used to generate a set of synthesized products that is unique to a bacterial species, for example.

(b) The polymerase chain reaction (PCR)

Isolation of DNA from tissues, cells, or viruses: The DNA is mechanically sheared during this procedure, generating large fragments.

DNA denaturation: The DNA is subjected to a high temperature so that it becomes single-stranded. Once it is single-stranded, it can act as a template for DNA synthesis.

Primer annealing: In order for DNA synthesis to begin, a primer (a short strand of DNA or RNA), must be base-paired to the template, and provide a 3′ hydroxyl at the end of its sequence. DNA polymerase can then link a series of nucleotides to the primer, one after another, all complementary to the template strand.

Since the primer must base-pair to the template, the particular sequence of the primer dictates where synthesis begins. The two primers depicted here base-pair at repetitive sequences on both strands. Repetitive sequences are short, highly conserved stretches of DNA that are present throughout the genomes of all bacteria tested so far. However, the distances between repetitive sequences differ from strain to strain.

DNA synthesis (primer extension): DNA polymerase adds nucleotides, complementary to the template.

Repeat denaturation, annealing, and DNA synthesis about 30 times.

Agarose gel electrophoresis: The PCR products are separated by size, generating a fingerprint that can be used to identify and compare bacterial strains.

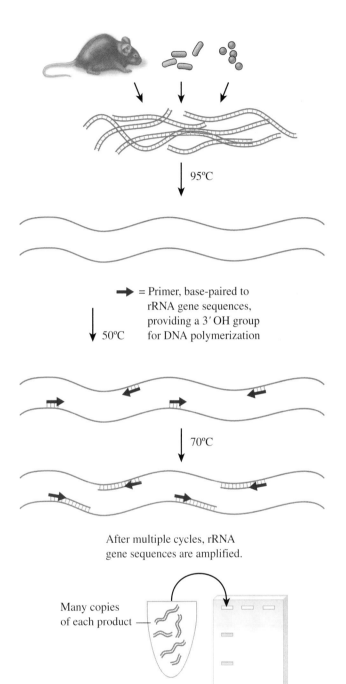

95°C

→ = Primer, base-paired to rRNA gene sequences, providing a 3′ OH group for DNA polymerization

50°C

70°C

After multiple cycles, rRNA gene sequences are amplified.

Many copies of each product

Materials

First Session: Restriction Digestion

Reagents
pBR322 containing viral genomes (prepared as knowns and as "clinical samples"). (These DNAs should be transformed into *E. coli* for storage and propagation, and isolated from the cells in preparation for this exercise.)
pBRSV SV40 virus DNA
pAM6 hepatitis B virus (HBV) DNA
pHPV-18 human papilloma virus (HPV) DNA
Restriction endonuclease, EcoRI prepared as a restriction mix (see table 33.2)

 All agents in red are BSL2 bacteria.

Equipment
Microcentrifuge
Vortexer
37°C heat block or water bath

Miscellaneous supplies
Laboratory marker
Latex gloves (when handling DNA samples)
Ice
1.5 ml microfuge tubes
Pasteur pipettes/bulb
1.0 ml serological pipette/pipettor
Micropipettors/tips (1–10 μl, 10–100 μl, 100–1,000 μl)

Second Session: Agarose Gel Electrophoresis and Staining

Reagents
Agarose
TBE: Tris-Borate-EDTA (108 g Tris-base, 55 g boric acid, 40 ml 0.5 M EDTA, pH 8.0, bring volume to 1 liter)
DNA sample loading buffer (tracking dyes): 0.25% bromphenol blue, 0.25% xylene cyanol, 30% glycerol in distilled water
DNA standard, lambda-HindIII, 1μg per 30 μl TBE; one per gel
DNA Blue InstaStain™

Equipment
Microwave oven
Horizontal gel electrophoresis system and power source

Miscellaneous supplies
Latex gloves (to protect DNA from deoxyribonucleases on hands)
Micropipettors/tips (1–10 μl, 10–100 μl)
125 ml Erlenmeyer flask
Bacterial waste beaker
Plastic ruler
Semilog paper

Procedure

First Session: Restriction Digestion

1. Pipette 7 μl of SV40 DNA into a microfuge tube, and place it on ice. Do the same for the hepatitis B DNA and the human papilloma virus DNA. In addition, obtain 7 μl of an unknown, "clinical sample," and place it on ice.

2. Add 23 μl of restriction enzyme mix to each of the four viral DNA tubes.

3. Using a different micropipette tip for each sample, mix well by gently pipetting up and down. If needed, centrifuge for a moment to bring the liquid to the bottom of the tube.

4. Incubate the samples at 37°C for at least 1 hour. They can be left longer, but should not be left overnight. After incubation, store the digested DNA in the refrigerator or freezer, or proceed to the next step.

Second Session: Agarose Gel Electrophoresis and Staining

1. Weigh out 0.4 g of agarose, and place it into a 125 ml Erlenmeyer flask. Add 50 ml of TBE to the flask, and swirl it gently. Using a lab marker, draw a line on the side of the flask indicating the level of fluid. Microwave it about 1 minute, checking to make sure it does not boil over. Return the flask to the microwave, and heat again as needed until there are no more flecks of agarose in the flask. If there has been obvious loss of volume through evaporation, add hot distilled water to the flask using the line you drew as a marker. Let the molten agarose cool until the flask is comfortable to handle, but still quite warm.

2. While the agarose is cooling, prepare the horizontal electrophoresis chamber according to the manufacturer's instructions (see figure 31.7).

3. When the molten agarose has cooled slightly, pour the gel and position the comb. With the long side of the electrophoresis chamber parallel to the edge of the lab bench, the comb should be positioned far to the left. It is important to keep in mind that the samples will run from the black lead end (the negatively charged cathode) toward the red lead end (the positively charged anode).

4. The agarose will solidify as it cools, within about 15 minutes. While the gel is solidifying, prepare your samples for loading. To each of your four samples, add 6 μl of sample loading buffer.

5. Each gel must also contain a DNA standard (see figure 33.5). Later you will use the standard to deduce the lengths of your restriction fragments. Add 6 μl of sample loading buffer to a 30 μl sample of standard.

6. When the gel is solid, gently remove the comb and the dams, and pour about 250 ml of TBE into the electrophoresis chamber until the gel is fully submerged.

7. Set a micropipettor at 36 μl. Load 36 μl of each sample into its designated well, changing the micropipette tip between samples. Load in this order:

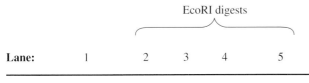

Lane:	1	2	3	4	5
Sample:	DNA standard λ-HindIII	SV40 DNA	HBV DNA	HPV DNA	Clinical DNA sample

8. Place the lid on the electrophoresis chamber, and connect the leads to the power source. Remember that the DNA will migrate from the black lead end toward the red lead end.

9. Set the power source at 80 volts (constant voltage), and allow the electrophoresis to proceed for 2 hours. As the gel runs, you will see that the tracking dyes are moving toward the red lead end as well. The dye fronts allow you to check the progress of the electrophoresis. The dye does not indicate the position of DNA fragments.

10. After 2 hours, turn off the power. Wearing gloves and using a spatula, gently remove the gel from the electrophoresis chamber. Place the gel onto a piece of plastic wrap, and stain the gel using the DNA Blue InstaStain method. Place a staining sheet over the gel, firmly running your fingers over the surface several times. Then place a glass or plastic plate on top of the gel with an empty beaker as a weight, and let the gel and staining sheet set for 15 minutes (see figure 31.9).

11. Remove the staining sheet, and place the gel into a shallow dish. Add distilled water heated to 37°C, changing the warm water every 10 minutes until the bands become visible. Gels can be left to destain overnight.

12. Using a plastic ruler, measure and record the distance migrated (cm) by each of the standard fragments (in the lamba-HindIII lane). Be sure to use the same start point for each measurement, such as the top end of the gel or the bottom of the well. Then measure and record in your laboratory report the distances migrated by your restriction digest fragments in each of the other lanes: SV40 DNA, HBV DNA, HPV DNA, and the "clinical sample."

13. Using a piece of semilog paper, graph the standard. Plot the distance migrated by each standard fragment on the x (linear) axis versus the log of its length (in base pairs) on the y (log) axis. When you use log paper, you do not need to calculate log. Alternatively, you may use a graphing program to plot the data.

14. Draw the best straight line. Do not include the data points from the largest two standard fragments (23,130 and 9,416). For an example of a semilog plot, see figure 33.6.

15. Using the distances you recorded for each of the restriction fragment bands, determine their lengths using the standard graph.

LABORATORY REPORT

NAME _____ DATE _____

LAB SECTION _____

Viral DNA Fingerprinting

1. Examine the restriction patterns of the three known samples and the single "clinical sample" and compare them. Can you identify the source of the unknown sample DNA? If so, what is it?

2. Complete the following table of DNA standard fragment lengths and migration distances based on your measurements.

Lambda-HindIII standard fragment lengths (base pairs)	Migration distance (cm)

3. Graph the standard fragment lengths versus migration distances using semilog paper or a graphing program.

4. List the migration distances of the band or bands you measured in each of the EcoRI digest lanes. Using the standard graph, deduce the size of each.

SV40 DNA		HBV DNA		HPV DNA		Clinical DNA sample	
Migration distance (cm)	Deduced length (bp)	Migration distance (cm)	Deduced length (bp)	Migration distance (cm)	Deduced length (bp)	Migration distance (cm)	Deduced length (bp)

5. Choose one of the DNAs from the table in question 4, and draw a restriction map of it. The map is circular, as in a plasmid restriction map. Include the following in the restriction map:

    -   the total length of the plasmid (in base pairs)

    -   the relative positions of the EcoRI

    -   the distance between these sites (in base pairs)

Isolation of Bacteriophage from Sewage and Determination of Phage Titer

Background

Virtually any type of cell is susceptible to virus infection; viruses cause disease in plants and animals, and can also infect procaryotes and unicellular eucaryotes. Viruses that infect procaryotes are known as *bacteriophages,* or *phages,* because when they were first discovered, they appeared to eat bacterial cells, generating a clearing, or *plaque,* on a lawn of susceptible bacteria. In reality, the bacteria are killed by **lysis** as newly produced phages are released from the damaged cells.

Like all viruses, bacteriophages consist of nucleic acid (RNA or DNA) surrounded by a protein coat, or capsid. Unlike some plant and animal viruses, bacteriophages are not enveloped. Some phages have elaborate structures for attaching to the bacterial surface and injecting nucleic acid into the cytoplasm. A diagram of one such bacteriophage, T4, is shown in figure 37.1.

Most bacteriophages are lytic; that is, each infection event leads to the production of new virions and the death of the cell by lysis. Some bacteriophages—most notably bacteriophage lambda (λ)—are categorized as temperate. Sometimes λ DNA is integrated into the bacterial chromosome, with its genes largely silent. The infected cell survives as a *lysogen.* In some λ infections, the DNA remains independent of the host chromosome, and is replicated many times over; its genes are expressed at high levels, and newly assembled phages are released. The "choice" between a lysogenic, nonproductive infection and a lytic, productive infection depends on environmental conditions. For example, UV exposure can cause a λ infection to switch from lysogenic to lytic. A typical lytic bacteriophage infection cycle is depicted in figure 37.2.

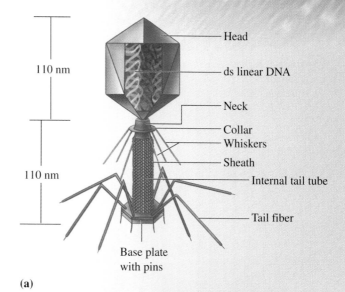

(a)

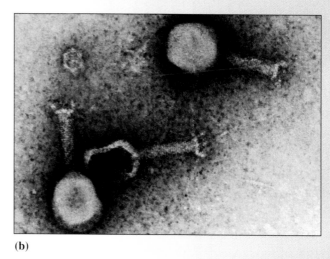

(b)

Figure 37.1 Bacteriophage or "phage" T4, a DNA virus of *E. coli.* (a) Diagram of phage T4. (b) Electron micrograph of phage T4 particles.

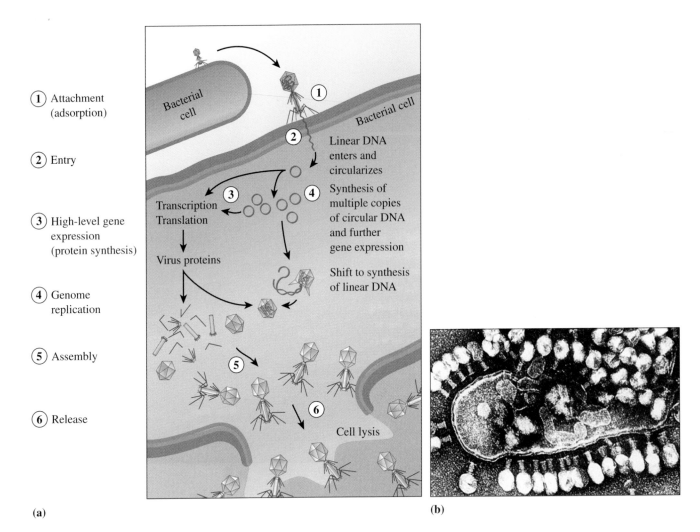

(a)

(b)

(1) Attachment (adsorption)

(2) Entry

(3) High-level gene expression (protein synthesis)

(4) Genome replication

(5) Assembly

(6) Release

Bacterial cell

Bacterial cell

(1)

(2) Linear DNA enters and circularizes

(3) Transcription Translation

(4) Synthesis of multiple copies of circular DNA and further gene expression

Virus proteins

Shift to synthesis of linear DNA

(5)

(6)

Cell lysis

Figure 37.2 The infection cycle of bacteriophage T4. (a) This kind of infection is "productive" because new viruses are produced. The steps listed on the left are generally applicable to any productive virus infection. (b) Electron micrograph of *E. coli* infected with phage T4 (36,500×).

In this exercise, we will focus on the bacteriophages of coliform bacteria. Coliform bacteria are relatively harmless microorganisms that live in large numbers in the intestines of mammals, where they aid in the digestion of food. *Escherichia coli* is a common fecal coliform bacterium. The presence of fecal coliform bacteria in water indicates that it has been contaminated with human or other animal feces, and that a potential health risk exists for those who use the water. Raw, untreated sewage contains large numbers of *E. coli*. Therefore, we will use raw sewage as a source of bacteriophages that infect *E. coli*.

In this exercise, you have the opportunity to: (1) amplify (increase the numbers of) phages in the sewage sample by allowing them to infect and reproduce within fresh *E. coli*, (2) collect the phages from the culture by centrifugation and filtration, and (3) detect and *titer* the amplified, isolated phages using a *plaque assay*. The assay is based on the fact that each plaque on a lawn of bacteria, although it contains 10^6 to 10^7 virions along with bacterial debris, represents a single infecting phage that entered one cell at the start of the culture. The infection then "spread" as the viruses reproduced and cells lysed, eventually forming a visible plaque (figure 37.3). The titer of a phage suspension, therefore, is determined by counting the number of plaques that form from a given volume of suspension. Phage titer is expressed as plaque-forming units (PFU) per milliliter (ml).

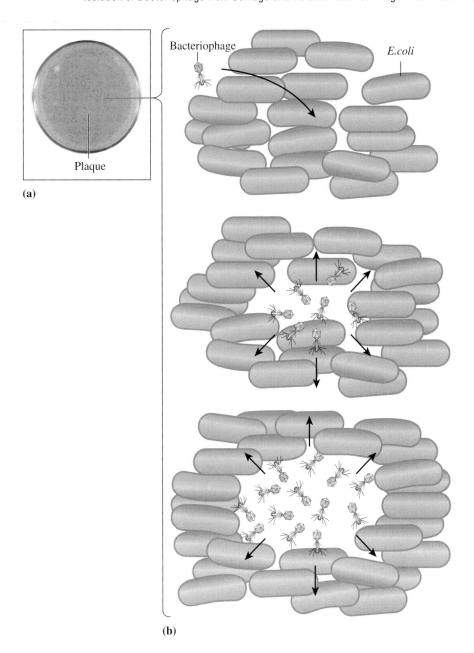

(a)

(b)

Figure 37.3 Phage plaques. (a) A lawn of *E. coli* B containing plaques. (b) Each clearing or plaque contains 10^6 to 10^7 bacteriophages and bacterial debris, but represents a single phage that infected one cell at the approximate center of that site.

Materials

Cultures
  Overnight culture of *E. coli* B
  1 (125 ml) Erlenmeyer flask containing 40 ml
    of raw sewage

Media
  Nutrient broth (1 g Peptone, 0.5 g yeast
    extract, 0.25 g NaCl, 0.8 g potassium
    phosphate, dibasic in 100 ml
    distilled water)
  10× strength nutrient broth (Peptone 20 g,
    yeast extract 10 g, NaCl 5 g, potassium
    phosphate-dibasic 16 g, in 200 ml
    distilled water)
  Warmed nutrient agar plates (6, 100 × 15 mm
    plates) (12–15 g agar/liter nutrient medium)
  Tubes containing 3 ml each of warm, top
    agarose (one per plate) (7.5 g agarose/liter
    nutrient broth, molten, cooled to 45°C)

Reagents
  Phosphate-buffered saline (PBS) (sodium
    chloride 1.6 g, potassium chloride
    0.04 g, sodium phosphate-dibasic 0.22 g,
    potassium phosphate-monobasic 0.04 g
    in 100 ml)

Equipment
  37°C incubator with shaker platform
  Water bath at 37°C
  Water bath at 45°C

Miscellaneous supplies
  5 ml pipettes/pipettor
  15 ml conical centrifuge tube
  Tube for collection and storage of
    phage filtrate
  Sterile 0.45 µm syringe tip filter
  10 ml syringe without needle
  1.5 ml microfuge tubes for preparing dilutions
  1.0 ml serological pipettes/pipettor or
    micropipettor/tips (100–1,000 µl)
  Laboratory marker

Procedure

Prior to today's lab, raw sewage was collected from a local sewage treatment plant. Yesterday, 50 ml of 1× nutrient broth was inoculated with *E. coli* B for overnight growth at 37°C with shaking.

First Session: Amplification of Bacterial Viruses

1. Pipette 5 ml of 10× nutrient broth into the flask containing 40 ml of raw sewage.
2. Inoculate the sewage in the flask with 5 ml of an overnight culture of *E. coli* B.
3. Inoculate a separate flask containing 45 ml of 1× nutrient broth with 5 ml of an overnight culture of *E. coli* B (one per class).
4. Incubate both cultures at 37°C, shaking for 24 hours.

Second Session: Bacteriophage Isolation and Plating:

Prior to today's lab, 2 ml of 1× nutrient broth was inoculated with *E. coli* B for overnight growth at 37°C with shaking. Earlier today, 100 ml of 1× nutrient broth was inoculated with a small volume of the overnight. This was done to obtain a culture in log growth by class time. *Note:* The instructor may choose to inoculate today's culture with the day-old "overnight" stored in the refrigerator.

1. Transfer 10 ml of the sewage-bacteria-bacteriophage culture into a centrifuge tube, and centrifuge the sample at 2,000 RPM for 5 minutes. Most of the remaining cells will be pelleted. The supernatant contains bacteriophage.

2. Prepare a 10 ml storage tube for the collection of bacteriophage supernatant as it is filtered. Then pipette the supernatant into a 10 ml syringe barrel fitted with a 0.45 micron filter. Gently slide the plunger, allowing the flow-through to drip into the storage tube. This step removes any remaining bacteria from the phage sample. The storage tube contains bacteriophage. It can be stored at 4°C and is stable for several months.

3. Prepare a series of microfuge tubes for making serial 10-fold dilutions of the bacteriophage suspension (performing the same dilution repeatedly in series is called serial dilution; see figure 37.4). Label six tubes 1–6. Into each tube, pipette 0.9 ml of sterile PBS.

4. ***Perform serial dilutions:*** Transfer 0.1 ml of phage suspension (that has been mixed well) into tube 1, and mix. Using the same pipette, transfer 0.1 ml of the sample from tube 1 into tube 2, and mix. Repeat this process, transferring 0.1 ml from tube 2 to tube 3, and so on, mixing each time, as shown in figure 37.4. Store the remaining phage suspension in the refrigerator.

5. Distribute 0.5 ml of log-phase *E. coli* into each of six microfuge tubes, labeled 1–6.

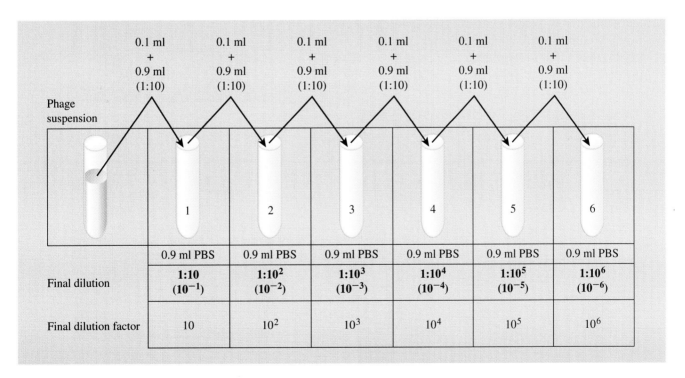

Figure 37.4 Serial dilutions of bacteriophage suspension. First, pipette 0.9 ml of PBS (diluent) into each dilution tube (numbered 1–6. Then transfer 0.1 ml of phage suspension in series, mixing each time.

6. To each tube of bacteria, add 0.1 ml of the corresponding phage dilution (0.1 ml of dilution 6 to cell tube 6, and so forth). *Note:* If you work from the most dilute to the least dilute, you can use the same pipette. Cap the tubes, and mix gently by inverting them.

7. Incubate at 37°C for 10 minutes to allow the phage to adsorb (attach) to the bacteria. This is your cell-phage mix.

8. In the meantime, label six warm, dry, nutrient agar plates 1–6 (one for each infection). Write on the bottom plate along the plate edge. Keep the plates in the 37°C incubator until you are ready to use them.

9. When you are ready to plate cell-phage mixes, collect your warmed, labeled plates from the incubator. Add the contents of cell-phage tube 1 to a vial containing 3 ml of top agarose (molten, at 45°C). Quickly cap the tube, and mix it by gently inverting it three times. Quickly pour the mixture onto warmed plate 1 (figure 37.5). You can tip the plate slightly to spread the top agarose. Push the plate aside, but do not pick it up until the agarose solidifies.

10. Repeat step 9 for each of the remaining five samples, 2–6.

11. Allow the plates to cool without being disturbed for approximately 10 minutes. When the top agarose has solidified, incubate the plates, inverted, at 37°C for 24 hours.

Third Session: Examination of Bacteriophage Plates, Phage Storage

1. Record the number of plaques on each plate in your laboratory report.

2. Using one of the least-crowded plates, pick an isolated plaque for long-term storage: Pipette 1 ml of PBS into a microfuge tube, and add 1 drop of chloroform. Then, using either the large or small end of a Pasteur pipette (depending on the size of the plaque and the space around it), pierce the agar surrounding the plaque, and pick out the agar "plug" containing the plaque (figure 37.6). Place the "plug," agar and all, into the 1 ml of PBS. The phage will diffuse into the PBS over time, and the chloroform will kill any remaining bacteria. Store the plaque in the refrigerator (4–10°C).

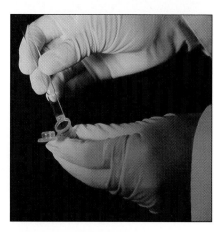

Figure 37.6 Picking a phage plaque for storage. Pierce the agar surrounding the plaque, and pick out the agar "plug" containing the plaque. Transfer the plug into a microfuge tube containing 1 ml of PBS and a drop of chloroform.

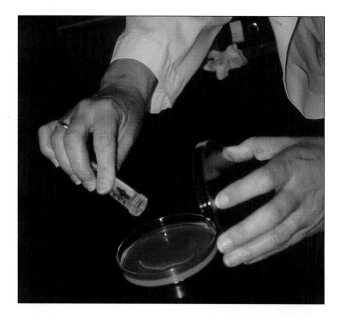

Figure 37.5 Plating phage. Once you have gently mixed the cell-phage–top agarose suspension by inverting it a few times, quickly pour the mixture onto a warmed agar plate.

LABORATORY REPORT

37

NAME _____ DATE _____

LAB SECTION _____

Isolation of Bacteriophage from Sewage and Determination of Phage Titer

1. Count the plaques on each plate. *Note:* If the plate is very crowded, it may be easier to count if you divide the plate in quarters or eighths and then multiply the count by 4 or 8, respectively. Then complete the following table.

Plate no.	Plaques per plate	Dilution factor	Volume of phage plated (ml)	Titer calculation (number of plaques) (DF) / volume plated (ml)	Titer: plaque-forming units (PFU) per ml

2. Do the results in the far right-hand column agree? Should they agree? What is the average titer of the amplified, filtered phage suspension?

3. Approximately how many bacteriophages are in the phage filtrate you collected?

4. A protocol calls for 10^9 phage particles as starting material. How much of your phage suspension would you need to have 10^9 phages?

5. Why is bacteriophage titer expressed as PFU/ml and not bacteriophages/ml?

6. Take a look at one of the phage plates, and comment on the plaques you see with respect to their appearance and dimensions. Do they all look alike? If two plaques differ in size or shape, what might that indicate about the bacteriophages in the two plaques?

7. What is the purpose of the amplification step?

8. You used a 0.45 μm filter to separate bacteriophages from any whole bacteria that remained after centrifugation. Why was this a proper choice of filter pore size? How big is an *E. coli* cell? How big is a typical bacteriophage? What else might be present in the bacteriophage filtrate?

9. You picked a single plaque from a phage plate for long-term storage. It is expected that all of the phages in the storage tube are identical. Why?

10. Describe and diagram how a bacteriophage plaque arises on a bacterial lawn.

11. Bacteriophage λ is a temperate phage. When λ is plated with susceptible *E. coli*, the plaques are visible but they are cloudy, not clear. Why are the plaques cloudy?

The Virus Infection Cycle: The One-step Growth Curve

Background

The steps described for the bacteriophage T4 infection cycle (see figure 37.2) are essentially the same for any type of virus that elicits a productive infection, no matter what the virus-host combination is. The molecular details of these steps (adsorption, entry, virus gene expression, viral genome replication, virion assembly, and release) are known for a number of bacterial, animal, and plant viruses.

In this exercise, you will observe the general features of a productive viral infection by completing what is known as a **one-step growth curve** experiment. *One step* refers to the fact that a single round of virus infection is assessed, and *growth* refers to the increase in numbers of virions that results from the round of infection. The experiment requires a *synchronous* culture—in this case, a uniform group of cells that are infected simultaneously with a uniform preparation of functional virions. In a synchronous culture the events in the infection cycle are expected to occur in each cell at nearly the same time. Therefore, the analysis of the whole culture over time reflects events occurring in a single cell.

The synchrony required for this kind of experiment has been achieved only with bacteriophages and their susceptible bacterial hosts in liquid culture. This is mainly because bacteriophages are much more efficient than plant or animal viruses at entering a cell once the virus has made contact with the cell. While the efficiency of infection for bacteriophages, expressed as a ratio of virions to infections, is 1:1 to 2:1 (for every one or two virions present, one is successful), the efficiency of plant and animal viruses ranges from 4:1 to 10,000:1!

Recall that during **adsorption and entry**, the virion attaches to a host cell, and its nucleic acid enters the cell. In the next phase of infection, the first steps of virus production occur: Virus genes are expressed, virus proteins are synthesized, and the viral genome is replicated. This phase is known as the **eclipse period**, because new virions are not yet formed; if the cells are taken from the culture and broken open chemically, no infectious virions are found. The eclipse period also

encompasses the bulk of the **latent period**, except that the latent period lasts through virus assembly, until the virions have formed and are released from the cell in the final phase, called **burst**. The term *burst* refers to the sudden increase in the number of free virions in the culture, not necessarily to the lysis that occurs in this—and in many but not all—productive virus infections. A representation of the one-step growth curve, based on the number of free virions in the culture over time, is shown in figure 38.1. The number of free virions at each time point is estimated by performing a plaque assay on a small sample of the liquid culture (figure 38.2).

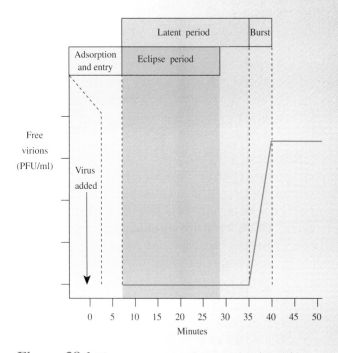

Figure 38.1 The one-step growth curve. A schematic representation of a virus infection cycle. The number of free virions at each time point is estimated by performing a plaque assay on a small sample of the culture. Keep in mind that although *eclipse* and *latent* are terms that convey lack of activity, there is much going on within the cell as virus genes are expressed at high levels (proteins are synthesized) and viral nucleic acid is replicated many times over.

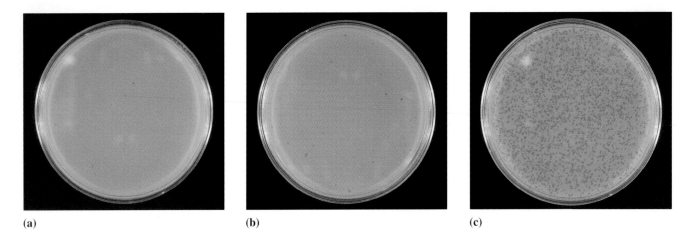

(a)　　　　　　　　　　　(b)　　　　　　　　　　　(c)

Figure 38.2 Result of plaque assays (phage plating) completed (a) 20 minutes, (b) 30 minutes, and (c) 40 minutes after initial infection. Each plate contains 0.1 ml of phage diluted 1:10,000 (10^{-4}).

Materials

Cultures
Overnight culture of *E. coli* B

Media
Nutrient broth (1 g Peptone, 0.5 g yeast extract, 0.25 g NaCl, 0.8 g potassium phosphate, dibasic in 100 ml distilled water)
Warmed nutrient agar plates (7, 100 × 15 mm plates) (12–15 g agar/liter nutrient medium)
Tubes (7) containing 3 ml each of warm, top agarose (7.5 g agarose/liter nutrient broth, molten, cooled to 45°C)

Reagents
T4 phage at 10^7 PFU per ml
or phage prepared in Exercise 37
Phosphate-buffered saline (PBS) (1.6 g sodium chloride, 0.04 g potassium chloride, 0.22 g sodium phosphate, dibasic, 0.04 g potassium phosphate, monobasic in 100 ml)

Equipment
37°C incubator with shaker platform (for overnight culture)
Water bath at 45°C
Water bath at 37°C

Miscellaneous supplies
15 ml conical centrifuge tubes
1.0 ml serological pipettes/pipettor or micropipettor/tips (100–1,000 µl)
Laboratory marker
Linear graph paper

Procedure

Prior to today's lab, 2 ml of nutrient broth was inoculated with *E. coli* B for overnight growth at 37°C with shaking. Earlier today, 100 ml of 1× nutrient broth was inoculated with a small volume of the overnight culture. This was done to obtain a culture in log growth by class time. *Note:* The instructor may choose to use the day-old overnight culture stored in the refrigerator.

1. Label seven warm, dry, nutrient agar plates with your name and the date, and label each with a time point: 20 minutes, 25 minutes, 30 minutes, 35 minutes, 40 minutes, 45 minutes, and 50 minutes. Keep the plates in the 37°C incubator until you are ready to use them.

2. Pipette 1 ml of the *E. coli* B culture into a sterile 15 ml conical centrifuge tube, and add 0.1 ml of bacteriophage (at about 10^7 PFU/ml: suspension commercially prepared or saved from Exercise 37). Mix well, and place the cap on the tube loosely.

3. Place the cell-phage mixture into the 37°C bath, and record the time. This is time zero. Make a note of what time it will be 19 minutes from now.

4. Incubate the mixture in the bath for 6 minutes. During this time, the bacteriophages adsorb to their host cells. Prepare the tubes for the dilutions you will do in the next step by labeling two 15 ml tubes #1 and #2 and pipetting 9.9 ml of sterile nutrient broth into each. Cap the tubes.

5. After the 6-minute incubation, centrifuge the cell-phage "adsorption" culture at 2,000 RPM for 5 minutes. Decant the supernatant into a waste receptacle, and resuspend the pelleted cells in 1 ml of fresh, sterile nutrient broth. Dilute the adsorption culture (infected cells) 10^4-fold by doing two 100-fold serial dilutions: Pipette 0.1 ml of the cells into dilution tube 1, mix well, and transfer 0.1 ml of cells from tube 1 to tube 2. Mix well.

6. Place the 10^4-fold dilution culture into the 37°C water bath. Check the time. When 19 minutes have elapsed since "time zero," collect your warmed, labeled nutrient agar plates, and go to step 7.

7. Add 2 drops of the remaining *E. coli* B culture (from step 2) to a tube containing 3 ml of top agarose (molten, at 45°C).

8. At *exactly* 20 minutes, transfer 0.1 ml of the 10^4-fold diluted culture to the tube of top agarose. Quickly cap the tube, and mix it gently by inverting it a few times; then immediately pour the mixture onto the warmed plate labeled "20 minutes" (see figure 37.5). You can tip the plate slightly to spread the inoculated top agarose.

9. Repeat step 8 at each of the subsequent time points: 25, 30, 35, 40, 45, and 50 minutes.

10. Once all the plates have cooled and the top agarose has solidified, incubate the plates, inverted, at 37°C overnight.

11. Count the plaques on each plate, and record the data in your laboratory report.

LABORATORY REPORT

38

NAME _____ DATE _____

LAB SECTION _____

The Virus Infection Cycle: The One-step Growth Curve

1. Count the plaques on each plate. Then complete the following table.

Plating time point	Number of plaques	Dilution factor	Volume of phage plated (ml)	Titer calculation (number of plaques) (DF) / volume plated (ml)	Titer: (PFU) per ml

2. Plot the number of PFU/ml versus time on the graph paper provided. Be sure to title the graph, label the axes, and include the units.

3. Using a different-colored ink, label the same graph with the phases of the virus infection cycle: adsorption and entry period, eclipse period, latent period, and burst.

4. If you continued to assay this infection culture beyond the 50-minute mark, to 100 minutes or so, what might the growth curve look like? Keep in mind that there are still plenty of cells remaining after the first infection cycle. Diagram and label a growth curve representing 100 minutes of phage-cell interaction, and briefly explain your answer. What would the growth curve look like if all the cells in the culture were infected in the first round of infection?

5. Step 2 of the procedure called for 0.1 ml of a phage sample at 10^7 PFU per ml. Approximately how many bacteriophages did you add to the *E. coli* B cells?

6. What was the titer of the bacteriophage you isolated in Exercise 37? How many milliliters of this suspension would you need to obtain the number of phages you calculated in question 5?

7. Why is it necessary to have a synchronous culture in order to formulate a one-step growth curve? Why is it impossible to generate a one-step growth curve using an animal virus?

8. In Exercises 37 and 38, bacteriophage suspensions were first diluted prior to plating with bacteria in top agarose. Why are dilutions necessary?

Infection of Plant Leaves with Tobacco Mosaic Virus

Background

The tobacco mosaic virus (TMV) is considered a prototype plant virus. In fact, it is perhaps the most studied and best understood of all the viruses. There are two reasons for its fame: First, TMV has potential impact on agriculture because of its wide range of hosts, including tobacco, tomato, and potato plants as well as ornamentals such as impatiens, geraniums, coleus, and African violets. Second, TMV is extremely stable—it is readily isolated from infected plant tissue and easily stored and maintained for laboratory studies.

TMV, a member of a large group of related viruses called *tobamoviruses*, is a rod-shaped, nonenveloped virus with a single-stranded RNA genome (figure 39.1*a*). The stability of TMV arises from tightly packed capsid proteins that make it resistant to conditions that would destroy most other types of viruses. As you can see in figure 39.1*b*, the tight association of TMV capsid proteins results in a rigid structure.

Plant viruses require help to breach the plant cell wall and gain access to the cytoplasm. This help comes in the form of prior tissue damage, from insects (in fact, some plant viruses are insect-transmitted) or from abrasions or wounds inflicted on the plant by weathering, machinery, or tools. Once in the cytoplasm, the virus replicates, and newly assembled TMV virions move throughout the plant, infecting most of its cells. The

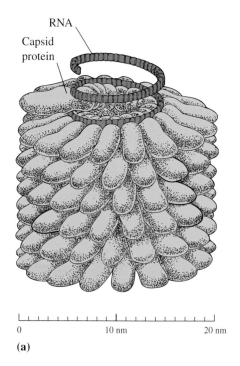

RNA

Capsid protein

0		10 nm		20 nm

(a)

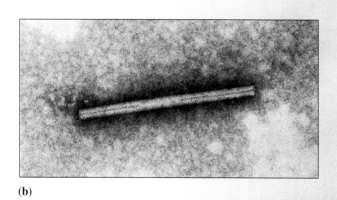

(b)

Figure 39.1 The structure of tobacco mosaic virus (TMV). (a) The capsid proteins are arranged in a helical array, tightly associated with each other and with the RNA genome. The RNA is said to be "positive sense" because it acts as a messenger RNA as soon as it enters the cell and associates with ribosomes. (b) Electron micrograph of TMV (400,000×).

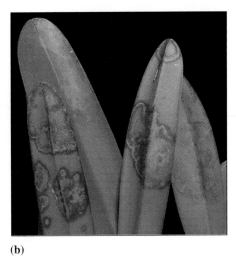

(a)   (b)

Figure 39.2 Viral lesions on plant leaves. (a) Tobacco mosaic virus (TMV) on *Nicotiana glutinosa.* (b) TMV infection of an orchid showing leaf color changes.

movement of virions to adjacent cells occurs through plasmodesmata, while movement to distal leaves or roots occurs via phloem. TMV infection stunts the growth of its host plant, and causes light and dark mottling (a mosaic pattern) on its leaves (figure 39.2).

In this exercise, you will extract TMV from dried tobacco leaves and detect the virus by applying it to leaves of a susceptible living plant such as coleus (figure 39.3).

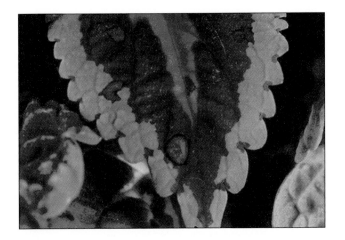

Figure 39.3 The effects of TMV on *Coleus blumei.*

Materials

Plants and reagents
 Young tomato plant (*Lycopersicon esculentum*) or coleus plant (*Coleus blumei*)
 1 g tobacco (from about 2 cigarettes)
 Positive and negative control inocula: TMV-infected and virus-free tobacco homogenates

Miscellaneous supplies
 Sharp knife, scissors, or razor blade
 Mortar and pestle
 Small piece of 600 grit sandpaper
 Small paintbrush
 Labeling tape
 Laboratory marker
 Small weighing dishes
 10 ml graduated cylinder or 10 ml pipette
 Cheesecloth
 Small funnel
 50 ml beaker

Procedure

1. Obtain two plants. Label both with the date and the names of your group members. Also label one plant "TMV inoculation"; it will be inoculated with the tobacco extract you prepare. The second plant will serve as either a negative control or a positive control. Check with your instructor to determine which control to use, and label the second plant accordingly.

2. If cigarettes are to be the source of tobacco, slit open two cigarettes using a sharp tool, and collect the tobacco in a small weighing dish.

3. Using a second weighing dish, weigh out 1 g of tobacco, and pour it into the mortar.

4. Add 10 ml of distilled water to the tobacco, and let it stand for 10 to 15 minutes.

5. Grind the tobacco for a few minutes with the mortar and pestle.

6. Separate the leaf extract (containing virions) from the tobacco remnants by filtration: Place two layers of cheesecloth into a small funnel positioned over a 50 ml beaker. Pour the contents of the mortar through the cheesecloth.

7. *Make a 1:10 dilution of the extract:* Pipette 0.9 ml of distilled water into a microfuge tube; then pipette 0.1 ml of extract from the beaker into tube 1, and mix well.

8. Choose one leaf on each of the plants (TMV-inoculated and control), and place a small piece of label tape around the stem of each.

9. Using sandpaper, *gently* scrape off the surface of a small area (about the size of a nickel) on each of the chosen leaves.

10. With the paintbrush, apply either the undiluted TMV suspension or the 1:10 dilution of tobacco extract to the scraped area on the leaf of the plant labeled "TMV-inoculated." Apply the prepared control homogenate to the second leaf. *Note:* This may be a positive control or a negative control. Be sure to record which you are using.

11. Place the plants in a greenhouse or other appropriate space, keeping the control plants away from the infected plants. **The virus can be transmitted from plant to plant if the leaves touch.** Observe the plant leaves every 2 or 3 days for 14 days and record your observations in your laboratory report.

LABORATORY REPORT

39

NAME _____ DATE _____

LAB SECTION _____

Infection of Plant Leaves with Tobacco Mosaic Virus

1. Record your observations at selected days after inoculation, commenting on the appearance of the infected leaf and the control leaves. If your group prepared a negative control plant, take a look at a positive control done by another group, and vice versa.

Days after inoculation	TMV-inoculated leaf	Positive control leaf	Negative control leaf

2. What conclusions do you draw, based on the observations recorded in question 1?

3. What was the purpose of rubbing the leaf with sandpaper prior to the infection?

4. A plant leaf lesion assay is similar to a plaque assay. In order to titer a suspension of TMV, a researcher spreads 0.1 ml of a 1:1,000 virus dilution evenly over the surface of a prepared leaf. After 3 weeks, she counts a total of 22 lesions on the leaf. What is the titer (here, expressed in infectious units [IU] per milliliter) of the original virus suspension?

5. The RNA genome of TMV is called "positive sense" because it acts as messenger RNA as soon as it enters the cytosol and associates with the host's translation machinery. Compare the tobacco mosaic virus and the banana streak badnavirus (a double-stranded DNA virus) with respect to gene expression.

Hematology and Serology

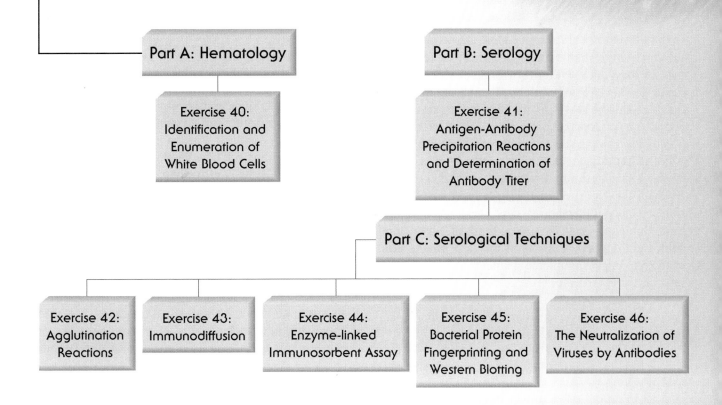

Part A: Hematology

Exercise 40:
Identification and
Enumeration of
White Blood Cells

Part B: Serology

Exercise 41:
Antigen-Antibody
Precipitation Reactions
and Determination of
Antibody Titer

Part C: Serological Techniques

Exercise 42:
Agglutination
Reactions

Exercise 43:
Immunodiffusion

Exercise 44:
Enzyme-linked
Immunosorbent Assay

Exercise 45:
Bacterial Protein
Fingerprinting and
Western Blotting

Exercise 46:
The Neutralization of
Viruses by Antibodies

40

Identification and Enumeration of White Blood Cells

Background

In the preceding sections, you have learned how protozoans, fungi, multicellular parasites, bacteria, and viruses reproduce within their hosts and sometimes cause disease. We now shift our focus from the infectious agent to the host—in particular, the human body—and how it can remain healthy even as disease-causing organisms gain access to it by way of air, food, and water. **Immunity**, the state of protection from infectious disease, is achieved by both nonspecific and specific mechanisms. Nonspecific immunity is provided by the skin and mucous membranes, which act as physical barriers to infection, and by several types of white blood cells (**leukocytes**), some of which are **phagocytic**, capable of engulfing microorganisms. Specific immunity, on the other hand, is conferred by **lymphocytes**, including leukocytes that bind specifically to foreign molecules (**antigens**) on the surfaces of invading organisms and infected body cells.

There are five major leukocyte types, each classified as either a **granulocyte** or an **agranulocyte**. The granulocytes (neutrophils, eosinophils, and basophils) contain cytoplasmic granules that are packed with degradative enzymes and mediators of inflammation. The agranulocytes consist of two morphologically and functionally distinct cell types (monocytes and lymphocytes) that have finer, less prominent granules. A white blood cell can usually be identified by the shape of its nucleus and by the presence or absence of granules. For example, a mature neutrophil has a distinctive multilobed nucleus and fine granules in its cytoplasm, while a lymphocyte has a rounded nucleus that fills much of the cell interior, and no obvious granules. The morphology, function, and population size for each leukocyte type is presented in table 40.1.

If you were to count all of the white blood cells in a particular volume of blood, you would be determining the *total* white blood cell count, a value expressed in number of cells per microliter (μl). If you were to identify each cell type as you count it, you would be generating a *differential* white blood cell count. A differential count is a measure of each leukocyte type (both mature and immature forms) and is likewise expressed as cells/μl for each cell type. As you will see, however, the most common cell in a sample of whole blood is the red blood cell, or **erythrocyte**. Typical erythrocyte counts range from 4,500,000 to 6,500,000 cells per μl, while total white blood cell counts range from 4,500 to 11,000 cells per μl.

The finger-stick method of blood collection provides a small amount of mixed capillary, arteriole, and venule blood. This type of blood collection is frequently used when only a small amount of blood is needed. It is also used on infants younger than 6 months of age, in young children, and in adults who have poor veins or whose veins cannot be used because of intravenous infusions. As with any blood collection or invasive procedure, all materials that touch the subject must be sterile (alcohol wipe, cotton ball, lancet, Band-Aid), and blood-contaminated materials must be disposed of properly. Review the **universal precautions**, in the laboratory safety section of this manual (see p. xiii) to learn the steps you must take when handling human source samples.

Although automated methods are now available for identifying and counting white blood cells, you will be identifying and counting cells with the aid of staining and light microscopy.

Table 40.1 The Morphologies and Functions of the White Blood Cell Types.

Blood cell type	Functions	Characteristic features (average diameter)	Number of cells/mm³ (µl) of blood percent of total WBC
Granulocytes			
Neutrophil	Important phagocytic cells in blood and tissues.	Multilobed nucleus, small cytoplasmic granules (10–14 µm)	3,000–7,000 cells/µl 35–71%
Eosinophil	Phagocytic cells that can migrate from the blood into tissues. Granule contents are particularly harmful to parasitic worms.	Bilobed nucleus, large cytoplasmic granules (10–14 µm)	100–400 cells/µl 0–4%
Basophil	Nonphagocytic cells with granules containing histamine and other compounds that act against parasitic worms. Basophils (in blood) and **mast cells**, a related cell type in tissues, also contribute to allergic and inflammatory responses.	Pinched U- or S-shaped nucleus, large cytoplasmic granules (10–12 µm)	20–50 cells/µl 0–2%
Agranulocytes			
Monocyte	Moderately phagocytic in the blood, these cells migrate into the tissues, becoming large, highly phagocytic cells called **macrophages**.	U- kidney-shaped nucleus, no visible granules (15–20 µm)	100–700 cells/µl 1–10%
Lymphocyte	B and T lymphocytes are present in the blood and in **lymphoid tissues** (the spleen and lymph nodes, for example). In response to contact with specific antigens, T cells develop into active killer cells, and B cells develop into antibody-secreting **plasma cells**.	Large, rounded nucleus with little visible cytoplasm (5–17 µm)	1,500–3,000 cells/µl 24–44%

Erythrocytes, or red blood cells, not shown here, are more numerous than white blood cell (4,500,000–6,500,000 cells/µl of blood. Erythrocytes are small (7–7.5 µm diameter), non-nucleated, and are biconcave in shape.

Materials

Blood Collection

Miscellaneous supplies
    Latex gloves
    Sterile, disposable safety lancets
      (Medi-Let®)
    Sterile cotton balls
    70% ethanol, or alcohol wipe
    Two clean microscope slides (one slide for
      sample, one spreader slide)
    Orange biological disposal bag (one per lab)
    Plastic sharps collector (one per lab)

Wright's Staining

Reagents
    Wright-Giemsa stain and buffer
    Distilled water
    70% ethanol

Equipment
    Light microscope

Miscellaneous supplies
    Staining support and drip pan
    Human blood film, smear
    Latex gloves

Procedure

Blood Collection

1. Use the middle of the outer segment of the third or fourth finger. To increase local blood flow prior to the puncture, you can wrap the finger in a warm, moist paper towel for 2–3 minutes.

> CAUTION: *Human Blood Handling Safety Note: Observe universal precautions (see page xiii); handle only your own sample.*

2. Clean the site with an alcohol wipe, and allow it to air-dry. Do not blow on the skin to dry the alcohol. Blowing can contaminate the site. It is important to completely air-dry the residual alcohol because it may cause rapid hemolysis when it contacts the blood.

3. Follow the Medi-Let® procedure to obtain your blood sample. Immediately dispose of the cartridge in the plastic sharps collector.

4. Wipe away the first drop of blood with the sterile cotton ball. The first drop of blood usually contains excess tissue fluid.

5. Place 2 drops of blood near one end of a clean slide. Take care not to touch your skin to the slide. Place the short end of a second, spreader slide into the blood drops, with a 30° to 40° angle between the two slides, until the blood spreads along the edge of the spreader slide (see figure 9.3). Just before the blood has spread completely along the edge of the spreader slide, push the spreader slide along the first slide to form a smear. The smear should show a gradual transition from thick to thin.

6. Label the slide with your name and the date, and let the smear air-dry.

7. Dispose of all materials properly.

Wright's Staining

1. Place slides on a staining support over a drip pan, and apply enough drops of Wright-Giemsa stain to cover the smear. **Count the number of drops you use.** Incubate for 1 minute at room temperature.

2. Add an equal number of drops of Wright's phosphate buffer (pH 6.4). Gently blow on the slide to mix the solutions, and incubate for 3–6 minutes at room temperature.

3. Rinse the slides with distilled water. Cleanse the back of the slide with 70% ethanol.

4. Allow the smear to air-dry.

5. *Examine the stained blood smear:* Scan the blood smear at 40× magnification. Erythrocytes (red blood cells) are rose to salmon in color, are biconcave, and have no nucleus. Most of the cells you see are erythrocytes. The leukocytes (white blood cells) are larger, nucleated cells, colored shades of blue by the stain. The nuclei of white cells will be blue to light purple; the cytoplasm will vary from pale pink (neutrophils), to pale gray (monocytes), to light blue (lymphocytes). Eosinophils and basophils are discernible by their granules (orange to rose for eosinophils; violet to blue for basophils). Neutrophils also contain granules that stain pink to purple.

6. *Perform a differential white blood cell count:* Shift the magnification to 100×. Identify and record the white blood cell types you see, starting from the sparse end and working toward the more dense end of the smear. Use the cross-sectional method of differential counting. Count from the bottom right of the sparse end; count up, count left, count down, count left, etc., as shown in figure 40.1. When you see a white blood cell, identify and record it in your laboratory report. Design a data table that will allow you to easily record your data. Stop when you have counted a total 100 white blood cells.

7. Calculate the percentage of each leukocyte type you counted. When counting 100 cells, the percentage is easily discerned; a count of 65 neutrophils means that neutrophils make up 65% of the total white cells in the sample.

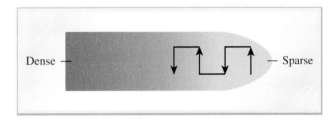

Figure 40.1 The cross-section method of counting white blood cells.

LABORATORY REPORT

NAME _____ DATE _____

LAB SECTION _____

Identification and Enumeration of White Blood Cells

1. Complete the following table based on your results.

White blood cell type	Labeled diagram of cell type	Number of cells counted	Percent of total cells	Typical percent range for cell type

2. Briefly state whether your results are within typical ranges for each cell type.

3. If you counted 100 total white cells over a portion of a smear that is equivalent to about 50 μl of blood, would that white blood cell count be within normal range?

4. Identify each of the following cell types. Give one reason for your answer in each case.

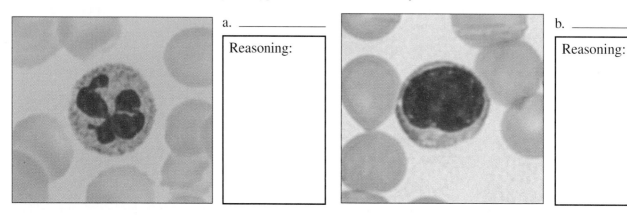

a. _____

Reasoning:

b. _____

Reasoning:

5. A 22-year-old female comes into the emergency room complaining of severe abdominal pain in the right lower quadrant. Her temperature is 39°C, and laboratory studies reveal a white blood cell count of 25,000/microliter.

   a. Is her total white blood cell count within normal range?

   b. She is diagnosed with appendicitis. One of our defenses against an infection such as this is phagocytosis. Thus, an important response to infection is the proliferation of phagocytic cells, in particular a white blood cell type that can be found in both the blood and in tissues and that has a multilobed nucleus. Given this, which of the five cell types might be most prominent in her blood?

6. White blood cells include those involved in nonspecific defenses such as phagocytosis and the release of histamine as well as those that operate in specific defense. Name the type of white blood cell that operates in specific defense. There are two subtypes in this category that function in particular ways in specific defense. Name the two cell subtypes, and state the role of each.

Agglutination Reactions: ABO Blood Typing

Background

As noted in Exercise 41, the highly specific and selective binding of antibodies to antigens has led to the development of a number of antibody-based diagnostic and research methods. In these methods (agglutination, precipitation, ELISA, Western blot, and others), antibodies are used as tools for the detection and quantification of drugs, hormones, and other molecules, as well as for the identification and characterization of viruses, cells, and tissues. Alternatively, such molecules, viruses, and cells (antigens) can be used to detect the presence of particular antibodies in a test sample. In the HIV screening test, for example, a person's serum is tested for the presence of antibodies specific for HIV, and not for the virus itself. In this case, the antigen is the tool, or the "known," and the antibody specificity is the "unknown."

When antibodies bind to cells, such as bacteria, yeast, or red blood cells, the cells clump together, or **agglutinate**. A visible *agglutination reaction* indicates that antibodies are binding specifically to cells, linking them together to form a large complex. So, just as antibodies bind to soluble molecules to form an insoluble precipitate, they bind to cell-bound molecules to form a clump of cells. Agglutination reactions are routinely used to type blood, to identify microorganisms, and to test serum samples for the presence of antibodies reactive against a particular microbe.

In blood typing, antibodies are used to detect red blood cell surface antigens such as those of the ABO system. These antigens consist of a core glycolipid called substance H. If substance H is modified by the addition of another sugar, N-acetylgalactosamine, it is an A antigen. If substance H instead has an attached galactose, it is a B antigen. If substance H stands alone, unmodified, it is neither A nor B, and is known as O (figure 42.1). A antigens are found on type A and type AB red blood cells, B antigens are found on type B and type AB red blood cells, and neither is found on type O red blood cells. Each of these blood types may be Rh^+ or Rh^-, depending on the presence or absence of another antigen, a protein called the Rh factor.

It may seem odd that normal molecules such as these are called antigens. In fact, all macromolecules are potential antigens, especially if they are transferred into a nonidentical person or to another animal through transfusion or transplantation. And, in autoimmune disorders, normal self-molecules such as these may be treated as foreign antigens. So, macromolecules can be called antigens because they have the potential to generate antibodies.

In this exercise, you will determine the blood type of either an aseptic blood sample or your own blood sample using the agglutination reaction. As shown in figure 42.2, agglutination, or *hemagglutination* (the term for red blood cell clumping), occurs when a blood sample is mixed with antibodies specific for its type. For example, if a sample of blood agglutinates when treated with antibodies to B but not when treated with antibodies to A or to the Rh factor, the sample is type B^-.

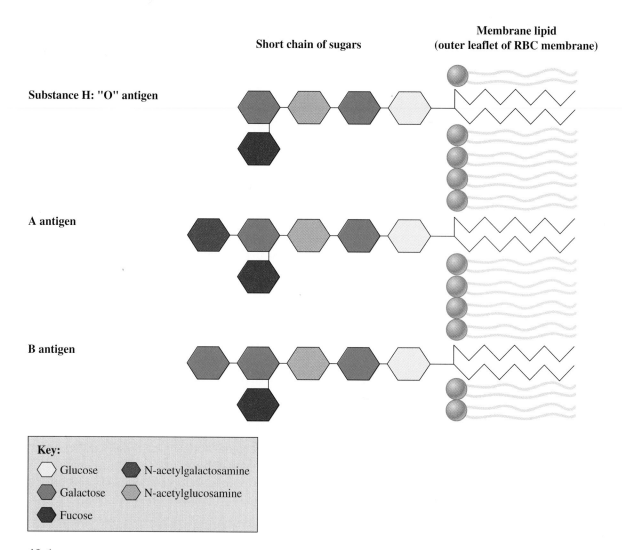

Figure 42.1 Illustration of the H (O), A, and B antigens. The antigens, which belong to a class of macromolecules called glycolipids, consist of a short chain of sugars covalently attached to membrane lipids.

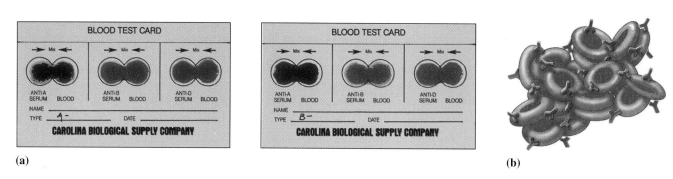

Figure 42.2 ABO-Rh blood typing with the antibody agglutination test. (a) *Left:* Type A⁻ blood reacts with antibodies to A (notice the clumped, particulate appearance of the samples) but not with antibodies to B or Rh (notice the smooth appearance). *Right:* Type B⁻ blood reacts only with antibodies to B. (b) A drawing of red blood cells agglutinated by antibodies.

Materials

Blood Collection

Miscellaneous supplies
Latex gloves
Sterile, disposable safety lancets
(Medi-Let®)
Sterile cotton balls
70% ethanol, or alcohol wipe
Two clean microscope slides (one slide for
sample, one spreader slide)
Orange biological disposal bag (one per lab)
Plastic sharps collector (one per lab)

Aseptic Blood and ABO-Rh Blood Typing

Reagents
Aseptic blood samples, anti-A, B, and Rh
antisera and materials provided in a Blood
Cell/Antisera BioKit
Miscellaneous supplies
Latex gloves

Procedure

Blood Collection

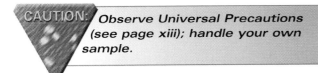
CAUTION: *Observe Universal Precautions (see page xiii); handle your own sample.*

1. Have a test card ready. Label it with your name and the date. If you are using an aseptic sample of blood, go to step 5. To prepare for the blood collection, consider using the middle of the outer segment of the third or fourth finger. To increase local blood flow prior to the puncture, you can wrap the finger in a warm, moist paper towel for 2–3 minutes.

2. Clean the site with an alcohol wipe, and allow it to air-dry. Do not blow on the skin to dry the alcohol. Blowing can contaminate the site. It is important to completely air-dry the residual alcohol because it may cause rapid hemolysis when it contacts the blood.

3. Follow the Medi-Let® procedure to obtain your blood sample. Immediately dispose of the cartridge in the plastic sharps collector.

4. Wipe away the first drop of blood with the sterile cotton ball. The first drop of blood usually contains excess tissue fluid. Allow the blood to drop onto each of the three areas on the test card marked "blood." Go to step 6.

ABO-Rh Blood Typing

5. If you are using a provided, aseptic blood sample, place 1 drop of the blood onto each of the three areas on the test card marked "blood."

6. Place 1 drop each of the anti-A serum, anti-B serum, and anti-D (Rh) serum onto the designated areas of the test card.

7. Mix each of the blood drop–antiserum sets, **using a fresh mixing stick for each sample.** Dispose of each mixing stick in the sharps container as soon as you are done with it.

8. Gently rock the test card for 1 minute, without allowing the samples to flow out of their designated test areas.

9. Examine the samples for agglutination. The anti-D reaction may take longer than the others. It may be easier to see agglutination if you tilt the card slightly.

LABORATORY REPORT

42

NAME _____ DATE _____

LAB SECTION _____

Agglutination Reactions: ABO Blood Typing

1. Record your agglutination results in columns 1–4 of the following table.

Sample I.D.	Anti-A reaction (+ or −)	Anti-B reaction (+ or −)	Anti- (Rh) reaction (+ or −)	Phenotype (blood type)	Genotype(s)

2. A person's ABO *phenotype* (A, B, AB, or O) arises from the expression of two alleles (alternative genes), one inherited from each parent. For example, the *genotype* of someone with type A blood may be either AO or AA.

   If we know the genotype of each parent, we can use a Punnett square to predict the possible genotypes of their offspring. Complete the following Punnett square, remembering that each parent contributes a single allele to each child. What is the genotype of their child with type A blood? What are the chances that they will have a child with type AB blood?

   Maternal
   ABO genotype

   B   O

   Paternal
   ABO genotype   A

   O

3. In the far-right-hand column of the table in question 1, indicate the genotype or possible genotypes for each phenotype you determined.

4. A child has type O blood, her mother is type A, and her father is type AB. Which parent could be a biological parent? Which parent cannot be a biological parent?

5. If a person is transfused with mismatched blood, an immediate *transfusion reaction* (an immune response against foreign blood group antigens) can occur. This happens because antibodies specific for foreign blood group antigens already exist in the recipient's blood. A person with type A blood, for example, has antibodies to the B antigen, even if he has never been exposed to type B blood. These antibodies arise in response to bacteria (normal flora) that have antigens very similar to the A and B antigens. Thus, this person with type A blood does not make antibodies to A-like bacterial antigens—the immune system considers these self—but does make antibodies to B-like bacterial antigens. Therefore, if a person with type A blood receives a transfusion of type B blood, the preexisting anti-B antibodies will induce an immediate and devastating transfusion reaction.

   For each of the following blood types, indicate whether the blood would also contain antibodies to A, antibodies to B, antibodies to both A and B, or no antibodies to A or B.

   Type A   _____

   Type B   _____

   Type AB   _____

   Type O   _____

6. A person with type O blood is considered a universal donor, while a person with type AB blood is a universal recipient. When type O blood is donated to a person with type A, AB, or B, packed cells are used rather than whole blood. Why?

7. Consider a sample of type AB⁺ blood. Each red blood cell in the sample has many copies of the A and B antigens on its surface but few copies of the Rh factor. Explain why this results in an anti-Rh agglutination reaction that is slower to form and less pronounced than either the anti-A or anti-B reaction.

8. A person's serum is mixed with *Salmonella typhimurium* cells on a slide. After a few minutes, particulates or clumps can be seen on the slide. What do you conclude about this result?

43

Immunodiffusion: Antigen-Antibody Precipitation Reactions in Gels

Background

As you observed in Exercise 41, when soluble antigen molecules become linked together by multiple antibodies, an insoluble precipitate forms. This precipitate, visible to the naked eye, can reveal the identity of an antigen (if the antibody specificity is known) or the specificity of the antibody (if the antigen is known). A precipitate also indicates that antibody and antigen molecules are present at optimal proportions for the formation of a large complex, or lattice. In this **equivalence zone,** there are about two to three antibody molecules for every one antigen molecule, leaving no free antigens or antibodies (figure 43.1).

In **immunodiffusion** tests, antibodies and/or soluble antigens are loaded into separate wells of a gel and are allowed to diffuse, each reagent moving radially into the gel. An immobile precipitate, visible as a band (precipitin line) in the gel, develops if specific antibody-antigen binding takes place, and if antibody and antigen components are present at optimal proportions. *Double immunodiffusion*, also known as Ouchterlony, is the most widely used gel precipitation technique in the research laboratory, while *radial immunodiffusion* and *immunoelectrophoresis* are principally used in clinical labs to test serum protein levels.

In double immunodiffusion, antigen and antibody preparations are loaded into separate wells of an agarose gel as shown in figure 43.2. In this example, the antibodies (specific for human serum proteins) are located in the center well, and the antigens (serum proteins) are located in the outer wells. Each substance diffuses from its well, and in time, white lines of insoluble precipitate appear at positions where antibodies have bound to their specific antigens at optimal proportions (the

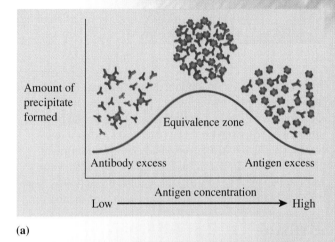

(a)

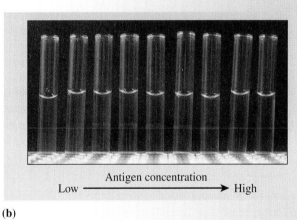

(b)

Figure 43.1 The precipitin curve. (a) The curve represents a series of antigen-antibody precipitin reactions showing that if either antibody or antigen is in excess, a complex does not form. The region of the curve in which complexes form is called the equivalence zone. (b) There is no visible precipitation in samples 1 and 2 because the antibody is in excess, and there is no precipitation in samples 8 and 9 because the antigen is in excess. Samples 3 to 7 represent the equivalence zone.

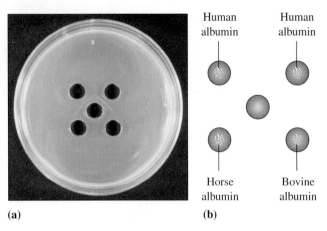

(a)   (b)

Figure 43.2 Double immunodiffusion assay. (a) Antibodies specific for human serum proteins were loaded into the center well, and antigens (human, bovine, and horse albumin) were loaded into the outer wells. After a few hours of diffusion, antigen-antibody complexes formed. The complexes do not diffuse, and they appear as white lines in the gel. The results indicate that the antibodies specific for human serum proteins bind with human albumin but not with horse or bovine albumin. (b) Diagram of the results in (a).

equivalence zone). In radial immunodiffusion, the antibodies are evenly distributed within a preformed 1% agarose gel. Therefore, in this test, only the antigen sample, loaded into a well, diffuses. As antigen molecules move through the gel, they bind to and carry antibodies, until the ratio of antigens to antibodies is optimal for complex formation. At this point, a ring of precipitation forms, and its diameter is proportional to the concentration of antigen loaded into the well; at higher concentrations, the diameter of the ring is greater because antigen molecules must migrate farther before they gather up enough antibodies to form a complex. Figure 43.3 shows the results of a radial immunodiffusion test for the human serum protein IgA (the antigen in this case).

The third type of immunodiffusion is called immunoelectrophoresis. Antigens are first loaded into wells of an agarose gel and are separated by charge in an elec-

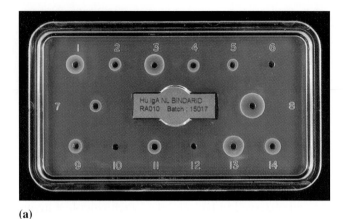

(a)

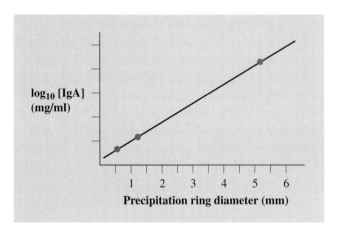

(b)

Figure 43.3 Radial immunodiffusion. (a) A photograph of a radial immunodiffusion gel. The agarose gel contains antibodies to immunoglobulin (Ig) A, one of the five classes of antibodies. Wells 7, 8, and 14 contain control IgA samples at 0.54 mg/ml, 5.4 mg/ml, and 1.3 mg/ml, respectively. Well 6, 10, and 14 contain no sample. (b) A graph of the log of the IgA concentration (mg/ml) versus the diameter of the precipitation ring using the standard samples in wells 7, 8, and 14 in (a).

trophoresis chamber. Antibodies are then used to detect the separated antigens; after being loaded into a trough that runs the length of the gel, they diffuse toward and complex with the antigens, and form visible lines of precipitate. Figure 43.4 shows the results of an immunoelectrophoresis analysis of human serum.

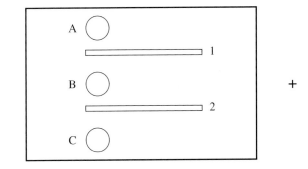

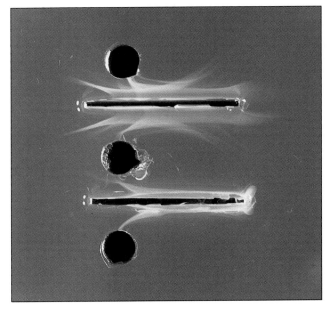

Figure 43.4 A photograph of an immunoelectrophoresis assay to detect serum proteins. Whole serum was loaded into wells A and B, and albumin was loaded into well C. The proteins migrate according to their net charges. For example, proteins that have a net negative charge migrate to the right, toward the positively charged anode. After electrophoresis, troughs 1 and 2 were both loaded with antibody specific for whole-serum proteins. The curved precipitin lines reveal the relative position of the major types of serum proteins.

Materials

Double Immunodiffusion (Ouchterlony)

Note: As an alternative to the following reagents and procedure, the Ouchterlony procedure can be accomplished using a kit (Edvotek #270).

Reagents
Agarose: 40 ml 1% (w/v) molten agarose in 0.05 M Tris-Cl, pH 8.6 (per pair)

Antibodies (table 43.1)
Serum antibody set (Carolina Biological Supply: #RG-20-2102)
Goat anti-bovine albumin
Goat anti-horse albumin
Goat anti-swine albumin

Antigens
Serum antigen set—bovine serum, horse serum, swine serum
Equipment
Microwave oven
Water bath at 55°C

Miscellaneous supplies
60 mm diameter petri dishes
Covered box for gel storage
Label tape
Laboratory marker
10 ml pipette/pipettor
Glass dropper (well cutter)
Micropipettor/tips (1–10 µl)

Radial immunodiffusion

Reagents
Human IgG, IgA, and IgM "NL" Bindarid™ radial immunodiffusion kit
Human serum

Miscellaneous supplies
Micropipettor/tips (1–10 µl)

Table 43.1 Sample Loading Order for Double Immunodiffusion Assays

	Center well*	Outer wells**	
Pattern A	Goat anti-bovine albumin	1. Bovine serum	
Pattern B	Goat anti-horse albumin	2. Horse serum	
Pattern C	Goat anti-swine albumin and goat anti-bovine albumin	3. Swine serum 4. Swine serum	

*The contents of the outer wells are the same for all three assays.
**The center well antibodies are different for each assay.

Immunoelectrophoresis

Reagents

    High-resolution electrophoresis buffer, pH 8.8

    1% agarose in high-resolution buffer, pH 8.8

    Antigens: bovine serum

        bovine albumin, 10 mg/ml

    Antibodies: anti-bovine albumin

        anti-bovine serum

Equipment

    Horizontal gel electrophoresis box

    Power supply

Miscellaneous supplies

    1 60 mm diameter petri dish

    Tape

    2 glass slides

    Glass or plastic dropper (well cutter)

    Micropipettor/tips (10–100 µl)

    Grade no. 1 Whatman paper or

        3MM paper

Procedure

Double Immunodiffusion (Ouchterlony)

1. ***Prepare 40 ml of 1% agarose:*** Add 0.4 g of agarose to 40 ml of 0.05 M Tris-Cl, pH 8.6, in a 125 ml flask. Microwave the mixture for about 30 seconds, checking to make sure it does not boil over. **Using a hot glove,** gently swirl the flask, and return it to the microwave. Heat for 15 seconds, repeating this until no flecks of agarose are visible in the flask. Let the molten agarose cool until the flask is comfortable to handle, but still warm.

2. Obtain three 60 mm diameter petri dishes. Writing with a lab marker on the plate bottom, label the three plates A, B, and C, respectively. Write your initials on all three plates. Pipette 5 ml of slightly cooled molten agarose into each dish. Allow the agarose to solidify, about 20 minutes.

3. Using the large end of a plastic or glass dropper (a diameter of about 0.5 cm), cut wells into each gel as shown in the following template. (See also figure 43.2.)

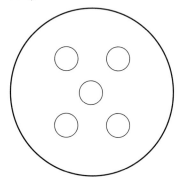

4. Label the outer wells 1, 2, 3, and 4 by writing on the plate bottom.

5. Changing micropipette tips between different reagents, pipette 20 µl of the appropriate antigen and 20 µl of antibody to the designated wells according to the loading order in table 43.1.

6. Line the bottom of the storage box with a moist paper towel, and place the dishes into the storage container. Make sure the dishes are level. Incubate the gels for 24 to 48 hours at room temperature to allow diffusion and banding. The gels can be stored in the refrigerator for several weeks if the box is kept moist.

Radial Immunodiffusion

1. Using a micropipettor, obtain 5 μl of human serum, and pipette it into a designated well of the RID assay gel.

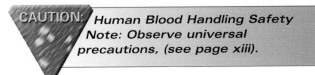
CAUTION: *Human Blood Handling Safety Note: Observe universal precautions, (see page xiii).*

2. Reserve three wells for standard concentrations: a high standard, a low standard, and a serum control (provided in the RID kit).

3. Place the gel into the moist box, and incubate for 24 hours at room temperature to allow diffusion and banding. Again, the gels can be stored in the refrigerator for several weeks if the box is kept moist.

4. *Read the results of your assay:* Measure the diameter of the circle of precipitate (in centimeters) for the sample you loaded and for the three standard samples. Record these results in your laboratory report.

Immunoelectrophoresis

1. *Prepare 40 ml of 1% agarose:* Add 0.4 g of agarose to 40 ml of high-resolution buffer, pH 8.8, in a 125 ml flask. Microwave the mixture for about 30 seconds, checking to make sure it does not boil over. **Using a hot glove,** gently swirl the flask, and return it to the microwave. Heat for 15 seconds, repeating this until no flecks of agarose are visible in the flask. Let the molten agarose cool until the flask is comfortable to handle, but still warm.

2. While the agarose cools, prepare a horizontal gel electrophoresis box by putting the dams securely in place. Also prepare a trough-forming apparatus: Obtain a 60 mm petri dish with lid, and tape a slide to each side as shown in figure 43.5.

3. Once the agarose has cooled so that the flask is comfortable to hold, pour the agarose into the unit until it completely covers the platform. Place the trough-forming apparatus at the center of the platform (see figure 43.5). Allow the agarose to solidify, about 10 minutes.

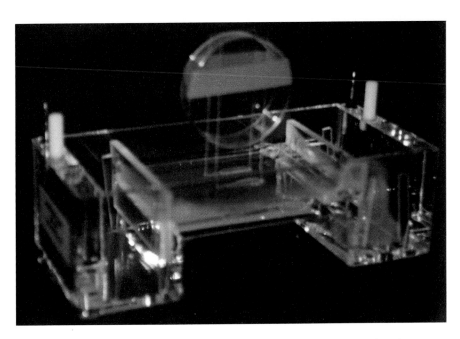

Figure 43.5 In preparation for immunoelectrophoresis, the trough-forming apparatus is placed at the center of the molten agarose.

4. When the gel is solid, gently remove the dams and the trough-forming apparatus. Using the large end of a plastic or glass dropper (a diameter of about 0.5 cm), cut wells into each gel as shown in the following template. (See figure 43.4.)

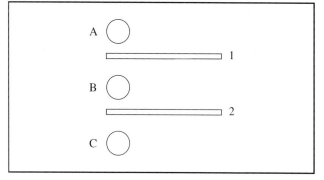

5. Pour high-resolution buffer, pH 8.8, into the electrophoresis box on either side of the gel, being careful not to pour onto the gel itself, 1 or 2 inches deep. Cut two pieces of Whatman chromatography paper wicks, and place them into the apparatus as shown in figure 43.6. Be sure that the paper is in contact with the gel and the buffer at both ends of the gel.

6. ***Load the antigens:*** Changing micropipette tips between samples, load 20 μl of bovine serum into wells A and C, and 20 μl of bovine albumin into well B. Do not load the troughs (figure 43.6*b*).

7. Electrophorese samples at 70 volts for 1.5 hours (figure 43.6*c*).

8. ***Load the antibodies:*** *After electrophoresis is complete,* load 50 μl of anti-bovine albumin into trough 1 and 50 μl of anti-bovine serum into trough 2. Again, change tips between samples.

9. Leave the gel in the electrophoresis apparatus, and wrap a moist paper towel and plastic wrap around it to create a moist container. Incubate for 24 to 48 hours at room temperature to allow diffusion and banding.

(a)

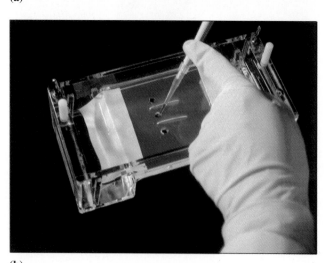

(b)

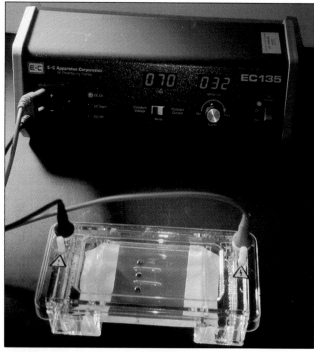

(c)

Figure 43.6 Immunoelectrophoresis. (a) The wicks must be in contact with the buffer and the gel. (b) Load antigens into the wells. Do not load the hentibodies into the troughs until electophoresis is complete. (c) Electrophorese the samples at 70 volts for 1.5 hours.

LABORATORY REPORT

NAME _____ DATE _____

LAB SECTION _____

Immunodiffusion: Antigen-Antibody Precipitation Reactions in Gels

Double Immunodiffusion (Ouchterlony)

1. A precipitin line represents a specific antibody-antigen reaction occurring between the antibodies diffusing from the center well with antigen diffusing from one of the outer wells. The precipitin line should be perpendicular to an imaginary straight line drawn from an outer well to the center well. The predicted results for pattern A are shown here. Predict the results for patterns B and C.

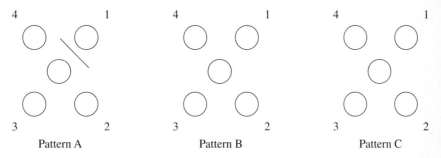

Pattern A   Pattern B   Pattern C

2. The double immunodiffusion assay also can be done to test the relatedness of antigens loaded into adjacent wells. If the two antigen samples are identical, a smooth corner forms where the two lines meet (identity). If the two antigen samples are not identical but related, then a spur forms at the corner (partial identity). Finally, if the two adjacent antigen samples are not related at all, two spurs form at the corner (nonidentity).

   Antibodies specific for all human serum proteins were loaded in the center well.

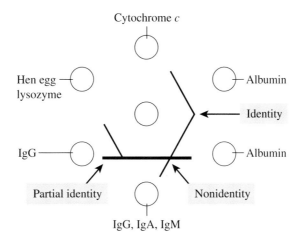

   No reaction is expected between the center well and the antigen wells containing hen egg lysozyme or cytochrome *c*.

3. Return to question 1, and consider whether spurs should be included in your predicted results. Add spurs to the diagram if they are expected.

4. Diagram your double immunodiffusion results. Do they agree with your predictions?

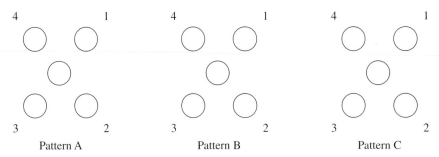

Pattern A   Pattern B   Pattern C

Radial Immunodiffusion

1. Record the three standard immunoglobulin (Ig) concentrations and the diameter of each resulting precipitin ring. Also record the diameter of the precipitin circle for the sample you loaded.

Ig standard concentration	Diameter of precipitate circle

Sample precipitin ring diameter: _____

2. *Graph the standard results: On semilog paper,* plot the three known, standard Ig concentrations (on the log scale) versus the diameter of the corresponding precipitin rings (on the linear scale). If you did not run standards, analyze the gel shown in figure 43.3a.

Comment on your results. The normal mean concentration of each antibody in serum is presented in table 43.2. The radial immunodiffusion test is often done to determine the concentrations of IgG, IgA, and IgM.

Table 43.2 The Five Classes or *Isotypes* of Antibodies (Immunoglobulins)	
Antibody isotype	**Mean concentration (mg/ml)**
IgG	13.5
IgA	3.5
IgM	1.5
IgD	0.03
IgE	0.0005

Source: *The Difco Manual.* Eleventh Edition. Difco Laboratories.

Immunoelectrophoresis

1. Diagram the results of the IEP assay.

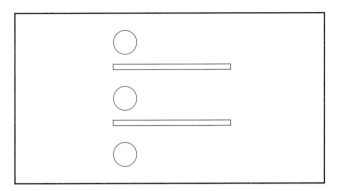

2. The following diagrams represent the results of two IEP assays done on the serum of a young child who has experienced frequent infections since infancy. What is your diagnosis?

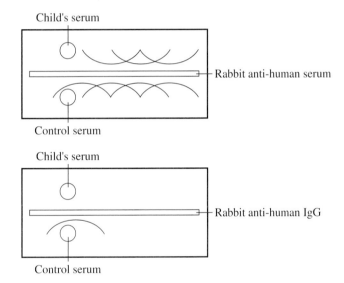

44

Enzyme-linked Immunosorbent Assay (ELISA)*

Background

In this laboratory experiment, we will simulate the transmission of a hypothetical infectious disease among members of the class. The infectious agent or antigen we are using is a harmless protein. However, for our purposes, consider it contagious and dangerous! The transmission of this hypothetical disease sets the stage for an ELISA (enzyme-linked immunosorbent assay), an antibody-based test that is commonly used as a research and diagnostic tool, and is the basis of the screening test for HIV. The ELISA takes advantage of the strong and specific attachment that occurs between an antibody and an antigen (thus the term *immunosorbent*). It is *enzyme-linked* because an enzyme is covalently attached to the tail portion of the antibody. The enzyme linked to the antibody is one that catalyzes the conversion of a colorless substrate into a colored product.

In an ELISA, the test sample, here simulated body fluid, is loaded into one well of a 96-well microtiter plate. Next, the enzyme-linked antibody specific for the infectious agent is added to each well. After washing to remove nonspecifically bound antibodies, the *chromogenic* (color-generating) substrate is added. The development of color in a well indicates a positive result; if the sample remains colorless, it is negative (figure 44.1). In addition, the intensity of color is an indication of the amount of reaction product in that well, which in turn correlates with the amount of enzyme-linked antibody—and so, the concentration of antigen—in the well.

Like the precipitin reactions (ID, IEP, and RID) and agglutination reactions, the ELISA takes advantage of a specific interaction between an antibody and an antigen, but unlike these, detection by ELISA doesn't require the formation of a large antibody-antigen complex. Therefore, the ELISA is much more sensitive than precipitin-type tests.

To carry out the experiment, each of you will be given a solution representing your own body fluid. You will exchange some of your body fluid with three other randomly chosen members of the class. Then you will perform an ELISA to test for the simulated disease agent in your "exposed" body fluid. Given the class results, it will be possible for you to trace the pathways of transmission and identify the original carrier or carriers of the disease.

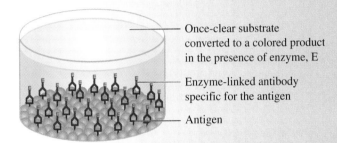

Once-clear substrate converted to a colored product in the presence of enzyme, E

Enzyme-linked antibody specific for the antigen

Antigen

(a)

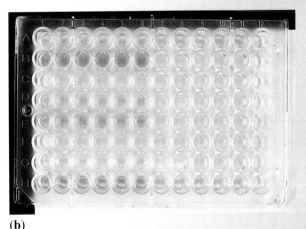

(b)

Figure 44.1 The enzyme-linked immunosorbent assay. (a) A diagram of the components of a typical ELISA depicting what would be a positive result. (b) A photograph of an ELISA plate after development. Row A1–6 contains a negative control, and row B1–6 contains a positive control. The test samples are in rows D (1–6), E (1–6), G (1–6), and H (1–6). No other wells on the plate contain test samples. The samples in rows D, E, and H are positive.

*Adapted from "Simulating the Spread of HIV" at *The Biology Project,* an interactive online resource for learning biology, developed at the University of Arizona: www.biology.arizona.edu

Materials

Exchange of "Body Fluids"

Reagents (see table 44.1)
  Two microfuge tubes that contain the solution
    representing body fluid (per person)
  1.0 ml serological pipette/pipettor or Pasteur
    pipette/bulb

Analysis of Samples by ELISA

Reagents (per pair)
  Positive control (contains infectious agent)
  Negative control (contains no infectious agent)
  Nonsharing fluid (partner A sample)
  Sharing fluid (partner A sample)
  Nonsharing fluid (partner B sample)
  Sharing fluid (partner B sample)
  Washing buffer
  Enzyme-linked antibody reagent
  Substrate (color-change reagent)

Miscellaneous supplies (per pair)
  96-well ELISA microtiter plate
  Micropipettor/tips (100 μl)
  10 ml pipette/pipettor

Pasteur pipettes/bulb
One piece of bench-coat absorbent paper

Procedure

Exchange of "Body Fluids"

1. Label the two body fluid tubes with your name,
   and place one of them (labeled "nonsharing") in
   the rack at the front of the room. Use the second
   body fluid tube (labeled "sharing") for the
   following steps.

   Proceed with steps 2 through 4 using your
   "sharing" tube. At the end of each exchange, you
   should have about the same volume of fluid you
   started with.

2. Using a transfer pipette, exchange about one-half
   of your sharing fluid with another person in the
   room. In table 44.2, record the name of the
   person you first made contact with.

   NOTE: Make sure you share with people in
   different parts of the room to prevent a local
   epidemic—spread "it" around.

Table 44.1 Table of Reagent Recipes for Simulated Infectious Disease Transmission

Simulated substance	Identity	Recipe
Body fluid	Sodium carbonate buffer	0.16 g sodium carbonate 0.27 g sodium bicarbonate in 100 ml distilled water
Viral antigen	Biotinylated albumin (Sigma-Aldrich #A 8549)	10 μl of biotinylated albumin at 6 mg/ml in 10 ml of sodium carbonate buffer
Wash buffer	PBS/0.1% Tween-20	32 g sodium chloride 0.8 g potassium chloride 4.48 g sodium phosphate, dibasic 0.8 g potassium phosphate, monobasic 2 ml Tween-20 Distilled water to 2 liters
Enzyme-linked antiviral antibody	Streptavidin peroxidase (Sigma-Aldrich #5512)	5 μl of streptavidin peroxidase (0.5 mg/ml 50% glycerol) in 50 ml of wash buffer
Substrate	TMB (tetramethylbenzidine) in phosphate citrate solution	Phosphate citrate solution: Combine 25.7 ml 0.2 M dibasic sodium phosphate and 24.3 ml 0.1 M citric acid solution with 50 ml distilled water.
	TMB tablets (Sigma-Aldrich #T 3405)	Dissolve 3 mg TMB in 30 ml of phosphate citrate solution. Add 5 μl 30% hydrogen peroxide. (Use same day; keep cold and dark.)

Table 44.2 ELISA Results and Potential Transmission Events

Name	Sharing ELISA results	Exchange 1	Exchange 2	Exchange 3	Analysis (excluded or not excluded as original carrier)	Non-sharing ELISA results

3. **At the instructor's signal,** find a different person to exchange one-half of your sharing fluid with. Record the name of your second contact.

4. **At the instructor's signal,** find a third person to exchange about one-half of your fluid with. Record the name of your third contact.

Analysis of Samples by ELISA

1. Join with a partner to proceed with the ELISA. The ELISA is designed to establish whether or not the infectious agent is present in your body fluid samples.

2. Write your names or initials on the plate edge. If you are using a Pasteur pipette to add samples and reagents to the ELISA wells, pipette 2 drops of each sample or reagent. **Always change pipettes between different samples and reagents.** If you are using a micropipettor to add samples and reagents, set it at 100 μl. **Always change tips between different samples and reagents.** Load controls and "body fluid" samples as shown in the following list and in figure 44.2. **Pipette carefully and accurately.**

To wells:	Add 2 drops or 100 μl of:
A1–A6 (row A)	negative control
B1–B6 (row B)	positive control
D1–D6 (row D)	nonsharing fluid (partner A)
E1–E6 (row E)	sharing fluid (partner A)
G1–G6 (row G)	nonsharing fluid (partner B)
H1–H6 (row H)	sharing fluid (partner B)

3. Incubate the samples at room temperature for 10 minutes, undisturbed.

4. Discard the liquid contents into the sink, and then place the plate facedown on absorbent paper with some force to remove any remaining liquid from the wells.

5. Using a 10 ml pipette, fill each of the wells that you used with wash buffer. Discard the wash solution into the sink.

6. Repeat step 5 twice (for a total of three washes).

7. Place the plate facedown on absorbent paper with some force to remove any remaining liquid from the wells.

8. Add 2 drops (or 100 μl) of enzyme-linked antibody to each of the wells (all the ones you used).

9. Incubate the samples at room temperature for 10 minutes, undisturbed.

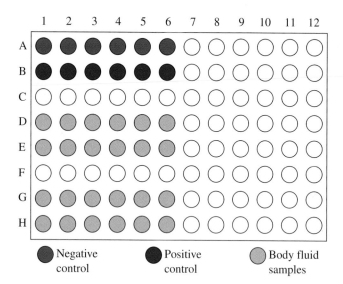

Figure 44.2 Diagram of a 96-well microtiter plate format and numbering system; colored areas represent loaded samples.

10. Wash the plate thoroughly by repeating steps 4–7.

11. Add 2 drops (or 100 μl) of substrate to each of the wells you used **Except those in rows D and G.** Rows D and G contain nonsharing fluids and will be assayed later.

12. After about 5 minutes of development, examine the qualitative results with respect to sample color changes.

Analyze Data; Determine Original Carrier(s)

1. Record your results in table 44.2 and on the board, providing your name, the results of the ELISA of your own "sharing sample" (+ or −), and the three people with whom you exchanged fluid, in order.

2. Join with two other pairs of students. As a group, work through the path of transmission to determine who the original carrier(s) might be.

3. Working in your original pairs, go back to your plate and add 2 drops (or 100 μl) of substrate to the wells in rows D and G (D1–6 and G1–6). Examine the sample results as in step 12. Remember that these are the original "nonsharing" fluid samples.

4. Record the results of the nonsharing fluid samples in table 44.2 and on the board. Determine if your group's conclusions were correct regarding the original carrier(s).

LABORATORY REPORT

NAME _____ DATE _____

LAB SECTION _____

Enzyme-linked Immunosorbent Assay (ELISA)

1. What happened in each test well? In the wells depicted below, diagram a positive ELISA and a negative ELISA. Include:

   • infectious agent (antigen)

   • enzyme-linked antibody where appropriate

   • the substrate and whether it is clear or colored

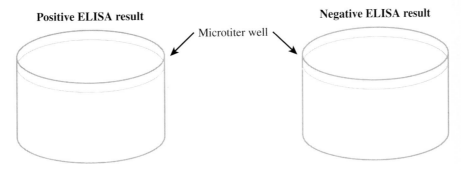

2. Describe the path of transmission that occurred in the class. In addition, formulate a flowchart that depicts the path.

3. It has been said that, "When you share fluid with someone, you are also sharing fluid with everyone they have previously shared fluid with." Addressing your results, discuss whether you agree or disagree with this statement.

4. In the ELISA, what would happen if you eliminated the washing step prior to adding substrate?

5. The ELISA screening test for HIV actually detects antibodies to HIV, not HIV itself, in a serum sample.

   a. Why does the presence of antibodies specific for HIV in serum indicate that a person is infected with the virus?

   b. In the HIV antibody ELISA, HIV antigens are first loaded into wells of a microtiter plate. The serum sample to be tested is then loaded into the well. If the serum contains antibodies specific for HIV, they will remain in the well, bound to HIV, even after the wash step. However, the antibodies are not enzyme-linked. A *second-step* antibody, specific for human antibody tails (IgG), is therefore added to the well next. This second antibody *is* enzyme-linked. An ELISA that requires two antibody steps is referred to as an **indirect ELISA**. Diagram a positive HIV antibody ELISA, including HIV, serum antibody, enzyme-linked second antibody, and substrate.

Bacterial Protein Fingerprinting and Western Blotting

Background

Just as any organism has a distinctive genome, it likewise has a **proteome**, a particular set of proteins characteristic of its species. The proteins can be extracted from cells and separated by gel electrophoresis to generate a pattern of bands, or a *protein fingerprint,* that is unique to that species. Although a typical bacterial cell contains about 2,000 different proteins, the fingerprint procedure outlined here reveals only those proteins present at high concentrations in the cells. Therefore, it will be possible for you to discern the fingerprint of each bacterial strain tested and compare it with others. It should be noted that different bacterial species have a number of proteins in common. However, these shared proteins will likely exhibit strain-specific differences in concentration, size (molecular weight), charge, shape, and reactivity to antibodies.

As you saw in Exercise 43, proteins separate by charge when exposed to an electric field. In order to separate proteins electrophoretically by size, they are first mixed with SDS (sodium dodecyl sulfate), a negatively charged detergent. SDS binds to all proteins in the mixture and denatures them so that each molecule assumes a random coil configuration—and becomes negatively charged. Thus, each protein will migrate toward the anode during electrophoresis, and its rate of migration will depend on its size. Larger random coil chains take longer to slither through the gel matrix (there is more drag), while smaller random coil chains migrate more rapidly through the gel matrix. Therefore, the mobility of each protein in an SDS-polyacrylamide gel (PAGE) is inversely proportional to the log of its molecular weight. The proteins in the gel are then visualized by staining, and the particular banding pattern, or fingerprint, of each bacterial strain can be discerned (figure 45.1).

While SDS-PAGE provides information about a protein's molecular weight—and here, a fingerprint—the identification of a specific protein within the proteome can be accomplished by following SDS-PAGE with **Western blotting**. In this method, proteins separated by SDS-PAGE are transferred from the gel onto the

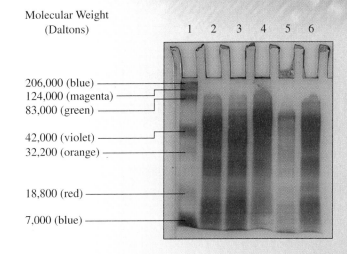

Molecular Weight (Daltons)

206,000 (blue)
124,000 (magenta)
83,000 (green)

42,000 (violet)
32,200 (orange)

18,800 (red)

7,000 (blue)

Figure 45.1 A stained gel containing bacterial proteins. A protein size marker (BioRad #161-0324) was loaded into lane 1. Proteins from *E. coli* B (lanes 2 and 6), *S. marcescens* (lane 3), *M. luteus* (lane 4), and *B. subtilis* (lane 5) were run on a 12% polyacrylamide gel, and the gel was stained with Coomassie blue. *E. coli* and *S. marcescens* are Gram-negative rods, *M. luteus* is a Gram-positive coccus, and *B. subtilis* is a Gram-positive rod.

surface of a membrane such as nitrocellulose or nylon. The membrane is then flooded with a solution containing labeled antibodies specific for a particular protein or proteins. In this exercise, the antibody label is a covalently linked enzyme, horseradish peroxidase, that converts a colorless substrate into a colored product. So, the Western is similar to an ELISA, except that the antigen is bound to a membrane instead of a plastic well, and a positive result appears as a colored band on the membrane instead of a colored liquid in a well. In this exercise, you will subject a duplicate, unstained fingerprint gel to Western blotting, and detect *E. coli* proteins with *E. coli*-specific horseradish peroxidase-linked antibodies (figure 45.2). At the completion of the Western blot procedure, the only bands you will see on the membrane will represent *E. coli* proteins. Keep in mind that some of these proteins (bands) may appear in non-*E. coli* samples, since different strains of bacteria have some proteins in common.

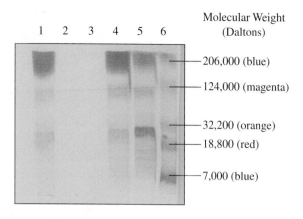

Figure 45.2 Photograph of a Western blot. Proteins were subjected to SDS-PAGE as described in figure 45.1, and transferred to a membrane for the detection of *E. coli* proteins with specific antibodies. The samples are a protein size marker (lane 1) and proteins from *E. coli* B (lane 2), *S. marcescens* (lanes 3 and 6), *M. luteus* (lane 4), and *B. subtilis* (lane 5). Note that antibodies have bound to proteins in the *E. coli* and *S. marcescens* samples (both Gram-negative), but not to proteins of *M. luteus* or *B. subtilis* (Gram-positive).

Materials

First Session: Preparation of Crude Protein Extracts

The following cultures and reagents are available in the Identification of Bacterial Protein Profiles Kit.

Cultures
   Bacterial strains grown as lawns on LB agar plates
   E. coli B, *Serratia marcescens*, *Micrococcus luteus*, *Bacillus subtilis*

Reagents
   Tris-EDTA-glucose (TEG) solution (25 mM Tris-Cl, pH 8.0, 50 mM glucose, 10 mM EDTA)
   TEG containing lysozyme 5 mg/ml, prepared the day of lab and stored cold
   Sample loading buffer (table 45.1)

Equipment
   Microcentrifuge
   37°C water bath or heat block
   Boiling water bath with microfuge tube rack

Miscellaneous supplies (for all parts of this exercise)
   Latex gloves

Laboratory marker
Micropipettor/tips (10–100 µl, 100–1,000 µl)
1 ml and 10 ml pipettes/pipettor
100 ml beaker
10 cc syringe, 18 g needle
1.5 ml microfuge tubes
Large weigh dishes for gel staining and Western blot incubations
Bench-coat absorbent paper
Receptacle to collect used antibody-blotto

Second Session: SDS-Polyacrylamide Gel Preparation and Electrophoresis

CAUTION: *Unpolymerized acrylamide is a neurotoxin. Always wear gloves and a lab coat when handling it. Since unpolymerized acrylamide may be present at the edges of polymerized gels, always handle gels with gloves.*

Reagents
   Precast 12% polyacrylamide gels
   Or, to cast gels: reagents for polyacrylamide gel formation
   Or, prepare gel solutions per table 45.1

Equipment
   Mini Protean II Cell
   Power supply
   Gel preparation kit (included in the BioRad Mini Protean II system)

Gel Staining/Western Blot Assembly and Transfer

Reagents
   Western transfer buffer (table 45.1)
   Blocking buffer (blotto) (table 45.1)

Equipment
   Power supply
   White light box
   Mini Transblot Electrophoresis Transfer Cell
   Optitran BA-S supported nitrocellulose membranes
   3MM chromatography paper

Table 45.1 Reagents for SDS-Polyacrylamide Gel Electrophoresis, Staining, and Western Blotting

SDS-Polyacrylamide Gel Electrophoresis

4× electrophoresis "running" buffer	Tris-base	90.85 g
	10% SDS	20 ml
	dH_2O	to 500 ml
	Adjust pH to 8.8 with HCl; store at 4°C.	
Sample loading buffer	4× running buffer	1.1 ml
	Glycerol	1.75 ml
	dH_2O	2.4 ml
	2-mercaptoethanol	0.5 ml
	0.1% bromphenol blue	0.25 ml
	10% SDS	4 ml
30% acrylamide:bisacrylamide 29:1	Acrylamide	60 g
	Bis-acrylamide	1.6 g
CAUTION: Wear gloves and goggles whenever handling acrylamide solutions; wear a mask when working with dry powder.	dH_2O	to 200 ml
	Filter through .45 μm filter; store at 4°C.	
10% sodium dodecyl sulfate (SDS)	SDS	50g in 500 ml dH_2O
Acrylamide gel solution (50 ml for 6, 12% gels)	30% acrylamide:bisacrylamide 29:1	20 ml
	4× running buffer	12.5 ml
	dH_2O	16 ml
	10% ammonium persulfate	1.5 ml
	Gently swirl to mix; when the pouring apparatuses have been prepared, add TEMED, swirl gently, and immediately pour the gels.	
	TEMED	25 μl
1× running buffer	4× running buffer	250 ml
	10% SDS	10 ml
Use this concentration for electrophoresis	dH_2O	to 1 liter

Gel staining (Do not stain a gel if you plan to proceed with Western blotting.)

Coomassie stain	Coomassie brilliant blue (250)	1.25 g
	Glacial acetic acid	50 ml
	Isopropanol	125 ml
	dH_2O	325 ml
Destain	Methanol	100 ml
	Glacial acetic acid	140 ml
Destain can be regenerated by running it through activated carbon in a filter funnel.	dH_2O	to 2 liters
Gel storage solution	dH_2O	425 ml
	Glacial acetic acid	50 ml
	Glycerol	25 ml

(Continued)

Table 45.1 Reagents for SDS-Polyacrylamide Gel Electrophoresis, Staining, and Western Blotting *(continued)*

Western transfer and blot development

Western transfer buffer	Methanol optima	200 ml
	Tris base	3.03 g
	Glycine	14.4 g
	dH₂0	to 1 liter, CHILL
Tris-buffered saline (TBS)	Tris-Cl	15.7 g
	NaCl	9 g
	pH	to 7.5
	dH₂0	to 1 liter
Blocking buffer (5% blotto)	Carnation instant nonfat dried milk 25 g in 500 ml TBS; store cold.	

Third Session: Membrane Treatment and Development

Reagents
Blocking buffer (blotto) (table 45.1)
TBS (table 45.1)
Antibody: rabbit anti-*E. coli* antigens, HRP-linked
Substrate: TMB 3,3,5,5,-tetramethylbenzidine

Procedure

First Session: Preparation of Bacterial Protein Extracts

1. Obtain a bacterial plate culture. The plate should have a confluent or nearly confluent lawn of bacteria (figure 45.3). Obtain a 10 ml test tube and a microfuge tube. Label each tube the same way the plate is labeled. Weigh the empty microfuge tube.

2. Pipette 4 ml Tris-EDTA-glucose onto the plate. Using a sterile inoculating loop or rubber policeman, gently scrape the entire plate to release the cells.

3. Transfer the fluid containing the released cells to the 10 ml tube, tilting the plate as needed to collect as much of the liquid and cells as possible. Once the liquid is in the 10 ml test tube, gently pipette up and down to disperse any cell clumps. You can also vortex the capped tube to suspend the cells.

4. Once there are no remaining clumps of cells, transfer 1 ml of the suspension into the

microfuge tube. Cap the tube, and centrifuge it for 1 minute at 14,000 RPM.

5. Decant the supernatant into a waste receptacle, and drain the remaining liquid onto a tissue. Weigh the tube once again to determine the weight of the cell pellet. Dispose of the tissue in a biohazard bag.

6. Resuspend the pellet with TEG so that the final concentration is 100 mg cells/ml. (For example, if you have 50 mg cells, suspend the cells in 0.5 ml TEG.) Mix the cells well by pipetting or vortexing, until there are no clumps.

7. Add one-tenth volume of TEG containing lysozyme. For example, if you have a 0.5 ml cell suspension, add 50 μl TEG-lysozyme. Mix the sample by pipetting up and down.

8. Incubate the sample at 37°C for 30 minutes.

9. Transfer 250 μl of your sample into a fresh, labeled microfuge tube, and add 750 μl of sample loading buffer to it. If you have 250 μl of sample or less, keep the sample in the original tube and add three times the volume of sample loading buffer to it. For example, if you have 130 μl, add 390 μl of sample loading buffer to the tube.

10. Cap the microfuge tube, and poke a small hole in the top using a needle. This will prevent the cap from popping open during the boiling step. Alternatively, use a screw-cap microfuge tube.

11. Place the capped tube into a boiling water bath for 10 minutes.

12. Allow the sample to cool, and then centrifuge it for 5 minutes at 14,000 RPM. Transfer most of

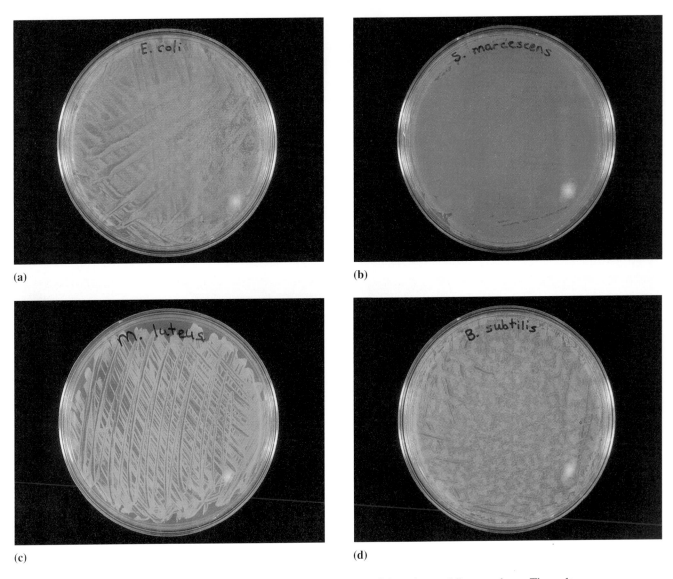

Figure 45.3 Confluent or nearly confluent growth of various bacterial strains on LB agar plates. The cultures were grown overnight at 30 °C. (a) *Escherichia coli*. (b) *Serratia marcescens*. (c) *Micrococcus luteus* (d) *Bacillus subtilis*.

the liquid into a fresh, labeled tube, leaving a small volume in the original tube. There may or may not be a visible pellet. Discard the tube containing the small volume along with any pellet.

13. Store the protein samples in the freezer, or proceed to the next step.

Second Session: SDS-Polyacrylamide Gel Preparation

1. If you are using precast gels, proceed to step 3. If you are using Reagents for Polyacrylamide Gel Formation (Edvotek #251), work on absorbent paper, wear gloves and a lab coat, and follow the instructions included in the kit; then proceed to step 3. To prepare the gel from scratch, go to step 2.

2. Working on absorbent paper and **wearing gloves and a lab coat,** pipette the appropriate volumes of 30% acrylamide:bis-acrylamide, 4× running buffer, distilled water, and 10% ammonium persulfate into a 100 ml beaker (see table 45.1). Do not add TEMED, until you are ready to pour the gel. When the gel-pouring apparatuses have been prepared, add TEMED and swirl gently. Using a 10 ml syringe fitted with a needle, aspirate about 8 ml of the gel solution. Place the bevel of the needle against the longer of the two plates, and push the plunger gently, allowing the solution to flow down between the plates. Place the comb, and allow the gel to polymerize (about 5 minutes). The BioRad gel apparatus and pouring procedure are shown in figure 45.4.

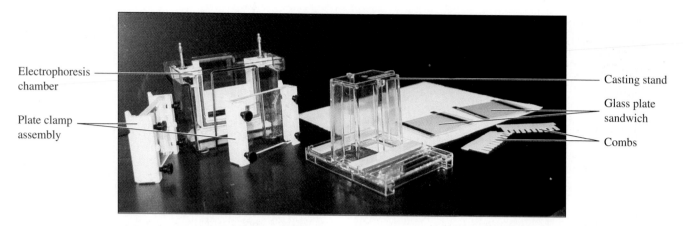

Electrophoresis chamber

Plate clamp assembly

Casting stand

Glass plate sandwich

Combs

(a) MiniProtean II electrophoresis cell components. The glass plate sandwiches consist of one long plate, one short plate, and two spacers.

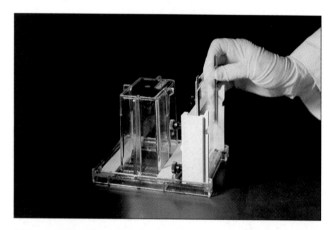

(b) Assemble the glass plate sandwich, and insert it into the clamp assembly. Place it onto the stage of the casting stand. Align the spacers and two plates so they are flush with the stage platform. Be sure that the spacers are straight and positioned at the outer edges of the plates. Tighten the two upper knobs slightly to hold the plates and spacers in place.

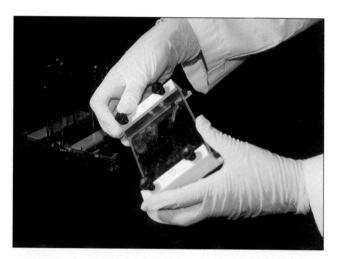

(c) Lift the clamp assembly/plate sandwich from the stage, and gently tighten all four knobs.

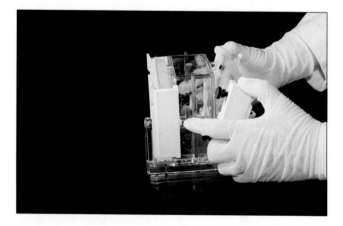

(d) Place the clamp assembly/plate sandwich onto the gasket of the casting stand, holding the assembly at an angle with its bottom end against the wall of the casting stand. Secure the assembly in place by pressing down on the white plastic clamp assembly (not the plates) and bringing the assembly upright beneath the plastic over-hang. The assembly should snap into position.

Figure 45.4 Assembly of the gel apparatus and preparation of the gel for SDS-polyacrylamide gel electrophoresis.

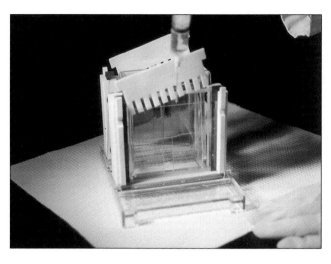

(e) Position the comb at an angle between the glass plates. Pour the prepared gel solution with a syringe fitted with a needle.

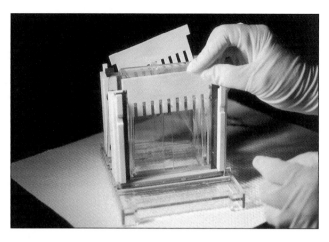

(f) Position the comb so that there is still space between teeth above the level of the short plate.

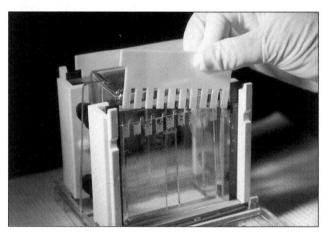

(g) Once the acrylamide has polymerized, gently remove the comb. Clean the long plate, above the level of the short plate, with a tissue. Dispose of the tissue in acrylamide waste.

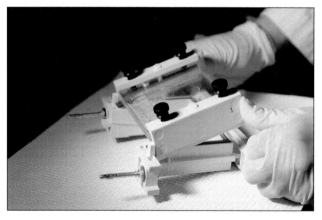

(h) Attach the clamp assembly/plate sandwich/gel to the cooling core. Slide the two wedges at the top of the clamp assembly into the two small slots in the cooling core, and snap the bottom of the clamp assembly into place.

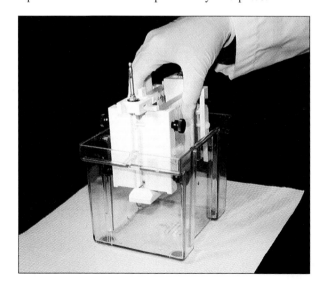

(i) Snap a second clamp assembly/plate sandwich/gel onto the cooling core, and place the entire assembly into the electrophoresis chamber.

3. If the protein samples you are working with were previously frozen, boil them for 10 minutes prior to loading the gel.

4. In consultation with the instructor, determine the appropriate loading for your particular gel. Record the loading order on a piece of notebook paper. The samples you load should include at least one "unknown" protein sample.

5. Load 20 µl of each sample into the designated lanes. *Note:* Change micropipette tips between samples. If possible, run each gel in duplicate, one for staining and protein fingerprinting, and one for Western blotting. Do not stain a gel if it is to be analyzed by Western blotting.

6. Run the gel at 70 volts for 1.5 hours.

For a Gel to be Coomassie-stained: Staining and Destaining the Gel

1. When SDS-PAGE is complete, place the gel into a large weigh dish containing Coomassie blue stain. Be sure that your gloves are wet with running buffer when you pick up the gel. Lift the gel by its two bottom corners.

2. Make sure that the gel is submerged in the stain, and cover the dish with plastic wrap. Place the dish on a rocker platform at a low setting for 1 hour. The gel can also be left overnight but will require more extensive destaining.

3. Place the gel into destain in a fresh weigh dish, cover the dish with plastic wrap, and return the gel to the rocker platform. After 5 minutes, pour the destain off into the proper receptacle and add fresh destain. Return the dish to the rocker platform for 1 hour, or until the destain solution is as blue as the gel itself. Repeat this until the blue protein bands begin to appear. Destaining can proceed overnight in a covered weigh dish.

4. When destaining is complete, soak the gel in storage solution (see table 45.1).

5. Place the gel on a light box, and examine the banding patterns. Record your observations in your laboratory report.

For a Gel to be Used for Western Blotting and Detection of *E. coli* Proteins

1. While the gels are running, prepare material for the transfer of proteins from the gel to the nitrocellulose membrane (Western transfer). Wearing clean gloves, cut one piece of the membrane to the dimensions of the gel (8 cm × 6 cm) and two pieces of Whatman paper (9 cm × 7 cm). Place these into a large weigh dish containing Western transfer buffer.

2. When SDS-PAGE is complete, place the gel into another large weigh dish containing Western transfer buffer.

3. Assemble the transfer apparatus. The assembly of the Western apparatus is shown in figure 45.5.

4. Place the assembled cassette into the transfer tank, and fill the tank with Western transfer buffer. Allow the transfer of proteins to proceed at 100 volts for 1 hour (or 30 volts overnight).

5. Disassemble the apparatus. The gel may be discarded, or it can be stained to confirm the transfer of proteins (protein bands should be absent or significantly weaker than those in the gel you stained).

6. Place the membrane, protein side up, into a large weigh dish, and pipette 20 ml of blotto over it. Be sure the membrane surface is covered with blotto. Blotto acts as a blocking agent because milk proteins bind to the membrane wherever proteins (here, bacterial proteins) are not already bound. Blocking is important to prevent the antibody from binding nonspecifically to the membrane in the next step. Incubate it at room temperature for 30 minutes with constant, gentle agitation on a rocking platform. The membrane can also be left in the blocking agent, covered with plastic wrap, and stored in the refrigerator for up to a week.

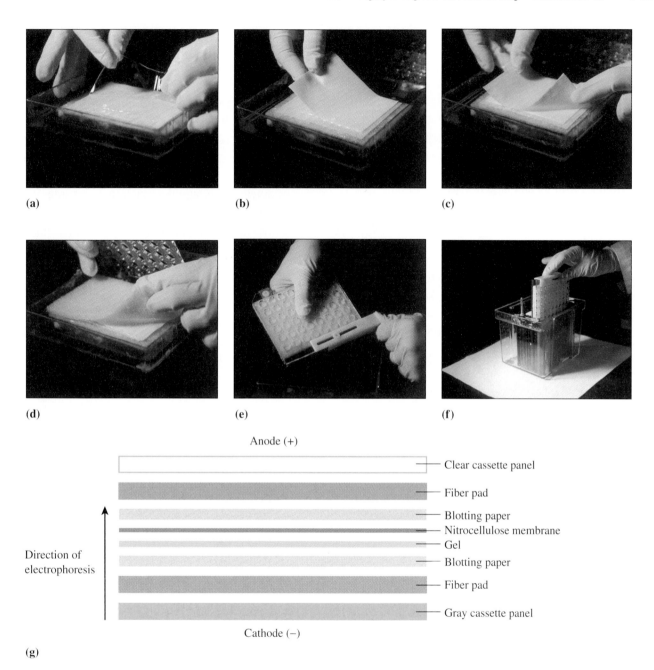

Figure 45.5 Preparation of the SDS-polyacrylamide gel for Western transfer. (a) The gel is soaked briefly in Western transfer buffer placed onto a piece of wet 3MM paper atop a fiber pad and gray cassette panel. On top of this are layered (b) nitrocellulose, (c) a second piece of wet 3MM paper, and (d) a second fiber pad. (e) The sandwich is then closed between cassette panels, and (f) placed into the electrophoresis chamber. (g) Diagram of assembly components.

Third Session: Membrane Treatment and Development

1. Holding the membrane with a gloved hand, pour the blotto into the sink, and pipette about 5 ml of blotto containing antibody (rabbit anti-*E. coli* antigens, HRP-linked, diluted as suggested by manufacturer) onto the membrane, completely covering it. Incubate at room temperature for 30 minutes with constant, gentle agitation on a rocker platform.

2. Holding the membrane with a gloved hand, pour the antibody-blotto into a receptacle (the antibody can be re-used). Wash the membrane three times with blotto: Pour about 50 ml blotto onto the membrane, and place it on the rocker platform for 10 minutes. Pour the blotto wash into the sink. Repeat this twice for a total of three washes.

3. Be sure the blotto has been fairly well drained from the dish. Pipette about 50 ml of TBS (without milk) onto the membrane, and incubate at room temperature on a rocker platform for 10 minutes. Pour the TBS off, and repeat the wash once with fresh TBS.

4. Pour off TBS, and pipette 5 ml of TMB substrate onto the membrane. The time necessary for color development will vary, usually from 3 to 10 minutes. Terminate the reaction by rinsing the blot with distilled water.

5. Allow the membrane to air-dry. Examine the banding patterns, and record your observations in your laboratory report.

LABORATORY REPORT

45

NAME _____ DATE _____

LAB SECTION _____

Bacterial Protein Fingerprinting and Western Blotting

Preparation of Crude Protein Extracts, Electrophoresis, and Gel Staining: Protein Fingerprinting

1. Compare the protein banding pattern in each lane of the stained gel. Provide a brief discussion of your results.

2. Given that the mobility of a protein in an SDS-polyacrylamide gel (PAGE) is inversely proportional to the log of its molecular weight, it is possible to determine the approximate size of an unknown protein or band using a standard graph. Measure the distance migrated by each band in the standard lane, and complete the following table.

Standard protein molecular weight (KD)	Distance migrated (cm)

3. Use these values to produce a graph on semilog paper. Graph the molecular weight of each standard protein (log scale) versus the distance migrated (linear scale).

4. Use the standard graph to determine the size of one of the proteins in the *E. coli* sample. Choose one of the strongest-staining bands. Once you have determined the approximate molecular weight of the *E. coli* protein, do some research on bacterial proteins to see if you can suggest the protein's identity.

Western Transfer and Membrane Treatment and Development: *E. coli* Protein Detection by Western Blotting

1. Which lane or lanes do you expect to be positive in the Western blot? Does this agree with your results?

2. Do any of the proteins in the other lanes appear on the Western blot? Briefly discuss this result.

3. What would have happened if you had eliminated the blocking step? What would your developed membrane have looked like?

The Neutralization of Viruses by Antibodies

Background

In Exercises 41–45, you learned about a number of serological methods, so called because each, in some way, takes advantage of antibody-antigen binding, usually employing antibodies to detect antigens. Let us now return to antibodies in their natural locale, circulating in the blood and tissue fluid, having been secreted from plasma cells (differentiated B cells) in response to foreign antigens. While it is obvious that these antibodies attach specifically to foreign substances, such as viruses, bacteria and bacterial toxins in the blood and tissue spaces, their effects upon these substances are not as apparent. How do antibodies help clear microbes, toxins, and foreign debris from the body?

As noted previously, antibodies tend to form complexes with their corresponding antigens, agglutinating cells and viruses, or precipitating free-floating molecules. These antibody-antigen complexes are at the core of three important infection-defeating mechanisms: neutralization, opsonization, and complement activation.

Neutralization is the simplest of the three mechanisms because it occurs simply as a consequence of anti-bodies binding to antigens. For example, antibodies bound to a bacterial toxin effectively block the toxin from contacting its target tissue and generating symptoms of the infection (figure 46.1a). Similarly, antibodies block, or *neutralize,* a virus by attaching to the molecules that the virus must use to attach to its host cell.

In **opsonization**, the tail portions of antibodies in the antibody-antigen complex become attached to receptors on macrophages, greatly enhancing the efficiency of phagoctyosis (figure 46.1b). In fact, a macrophage takes in and destroys a substance about 4,000 times faster when it is coated with antibodies. **Complement activation** can lead to the death of bacterial cells (mainly Gram-negative cells) and the destruction of enveloped viruses (figure 46.1c). Complement is a set of about 20 serum proteins that act in a cascade of steps. The cascade begins with the binding of the first component, C1, to two adjacent antibodies (class IgM or IgG) that are in turn attached to antigen on the surface of a bacterium; the cascade ends with the formation of a large pore in the bacterial membrane. With this loss of membrane integrity, fluid rushes into the bacterial cell, which then bursts, or *lyses.*

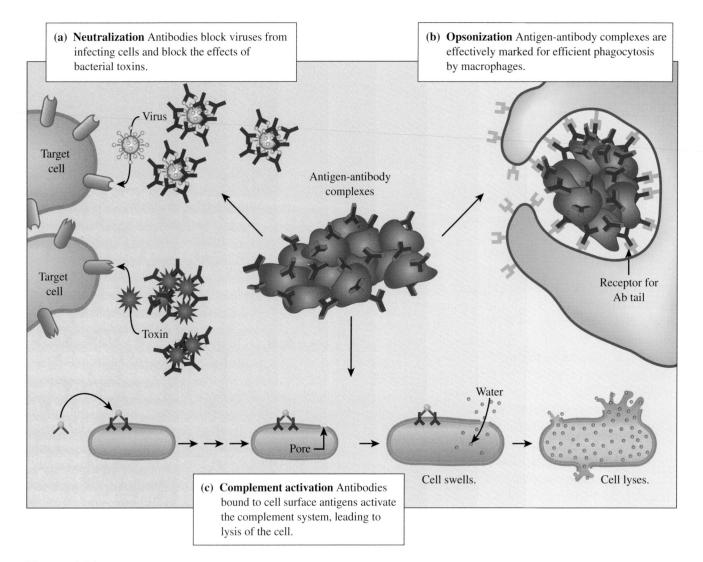

(a) Neutralization Antibodies block viruses from infecting cells and block the effects of bacterial toxins.

(b) Opsonization Antigen-antibody complexes are effectively marked for efficient phagocytosis by macrophages.

Virus

Target cell

Target cell

Toxin

Antigen-antibody complexes

Receptor for Ab tail

Water

Pore

Cell swells.

Cell lyses.

(c) Complement activation Antibodies bound to cell surface antigens activate the complement system, leading to lysis of the cell.

Figure 46.1 Antibody-mediated mechanisms of antigen disposal. The binding of antibody to foreign cells and molecules results in agglutination, the formation of large complexes. Antibodies mark these substances for defeat by (a) neutralization, (b) opsonization, and (c) complement activation.

In this exercise, you will observe the impact of antivirus antibodies on the capacity of viruses to infect cells. The virus in this case is bacteriophage T4, and the susceptible host cell is *E. coli* B (figure 46.2). Keep in mind that, although bacteria have ways of defeating virus infections (namely, restriction endonucleases), they do not exhibit specific immunity, and they do not make antibodies. For our purposes, however, consider the bacteriophage a pathogenic human virus, and the antibodies a result of a specific immune response to the virus.

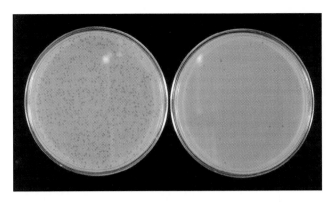

Figure 46.2 The effects of antibodies on virus infectivity. On the left is a photograph of bacteriophage T4-*E. coli* plating in the absence of antibody treatment. On the right is a photograph of bacteriophage T4-*E. coli* plating after the phage suspension was incubated with antibodies to T4 for 8 minutes. Viral neutralization is apparent in the diminished number of plaques on the plate on the right.

Materials

Reagents
  Disease Prevention Kit

Equipment
  37°C incubator with shaker platform
  Microwave oven or Bunsen burner
  Water bath at 48°C

Miscellaneous supplies
  Laboratory marker
  100 ml beaker
  Test tube rack

Procedure

First Session

Prepare Overnight Culture of *E. coli* B, and Label Dilution Tubes, Soft Agar Tubes, and Agar Plates

1. Inoculate 5 ml culture broth with *E. coli* B. Incubate the culture overnight at 37°C with shaking.
2. Label the 17 dilution broth tubes as shown:

   P-1, P-2, P-3, P-4, P-5, P-6, A-1, AP-2, 2AP-3, 2AP-4, 2AP-5, 4AP-3, 4AP-4, 4AP-5, 8AP-3, 8AP-4, 8AP-5

- P stands for **P**hage (T4 bacteriophage), and A stands for **A**ntiserum (antibody to bacteriophage T4).
- The negative number to the right indicates the dilution of the sample: −1 means 10^{-1}, or a 1:10 dilution; −2 means 10^{-2}, or a 1:100 dilution, and so forth.
- The number to the left of the letters A/P refers to the time (in minutes) that the antiserum and phage will be allowed to interact prior to plating.

3. Place the AP-2 tube into the 37°C incubator, and leave the others at room temperature.
4. Label the caps of the six tubes of soft agar: 1, 2, 3, 4, 5, X.
5. Label six agar plates: P-5, P-6, 2AP-5, 4AP-5, 8AP-5, 8AP-5X.

Second Session:

Dilute Stocks of Phage and Antiserum

1. Measure 1.1 ml of the T4 phage stock, and transfer 0.1 ml into the tube labeled AP-2 located in the 37°C incubator and 1 ml into the tube labeled P-1. Cap the tubes, and mix well. Return the AP-2 tube to the incubator.
2. Pour the entire contents of the anti-T4 antiserum container into the tube labeled A-1. (The antiserum is now diluted 1:10 [10^{-1}]. Cap the tube, and mix well.

Prepare Soft Agars, and Prepare and Plate Phage Dilutions (P-1 to P-6)

3. Place the six tubes of soft agar into a 100 ml beaker containing about 2 inches of water. Boil the water using a microwave oven or a Bunsen burner to melt the soft agar. Once the soft agar is molten, let it cool until you can touch it, but it is still quite warm (about 50°C). Place the tubes into the 48°C water bath. Once the tubes have been at 48°C for about 10 minutes, loosen each cap, but keep the contents covered.
4. Transfer 0.1 ml of the overnight culture of bacteria into each of the soft agar tubes *except* tube X. You may use the same pipette to inoculate tubes 1, 2, 3, 4, and 5.

5. Complete the serial dilutions of phage (P-2 to P-6), using a fresh pipette for each of the following transfers: Transfer 1 ml of the P-1 mixture (from step 1) into tube P-2 , cap the tube, and mix by shaking; transfer 1 ml from P-2 into P-3, cap it, and mix by shaking; transfer 1 ml from P-3 into P-4, cap it, and mix by shaking; transfer 1 ml from P-4 into P-5, cap it, and mix by shaking; transfer 1 ml from P-5 into P-6, cap it, and mix by shaking.

6. If the molten soft agar tubes have been inoculated with bacteria (step 4), transfer 1 ml of phage from dilution tube P-5 into soft agar tube 1. Quickly cap the soft agar tube, and gently invert it a few times to mix. Immediately pour the contents of the tube onto the agar in the plate labeled P-5 (see figure 37.5). Cover the plate, and tilt it slightly to spread the soft agar evenly. Allow the soft agarose to solidify (about 10 minutes).

7. Repeat step 6, transferring 1 ml of phage from dilution tube P-6 into soft agar tube 2, and plating the phage as before.

Prepare Antiserum-Phage Reaction; Dilute and Plate the Reaction at 2, 4, and 8 Minutes

8. Transfer 1 ml of antiserum from tube A-1 (from step 2) into tube AP-2 (from step 1) taken from the 37°C incubator. Cap the tube, and mix by shaking. Record the exact time, or set a timer for 8 minutes. Three teams (A, B, and C) should be ready to plate phage (team A at 2 minutes: steps 9 and 10; team B at 4 minutes: steps 11 and 12; and team C at 8 minutes: steps 13 and 14).

Team A

9. At exactly 2 minutes, quickly prepare serial dilutions (AP-2 to 2AP-5), using a fresh pipette for each of the following transfers: Transfer 1 ml of the AP-2 mixture into tube 2AP-3, cap the tube, and mix by shaking; transfer 1 ml from 2AP-3 into 2AP-4, cap it, and mix by shaking; transfer 1 ml from 2AP-4 into 2AP-5, cap it, and mix by shaking.

10. Transfer 1 ml of sample from the dilution tube 2AP-5 into the inoculated molten soft agar tube 3. As in step 6, quickly cap the soft agar tube, and gently invert it a few times to mix it. Then immediately pour the contents of the tube onto the agar plate labeled 2AP-5 as described in step 6.

Team B

11. At exactly 4 minutes, quickly prepare serial dilutions (AP-2 to 4AP-5), using a fresh pipette for each of the following transfers: Transfer 1 ml of the AP-2 mixture into tube 4AP-3, cap the tube, and mix by shaking; transfer 1 ml from tube 4AP-3 into tube 4AP-4, cap it, and mix by shaking; transfer 1 ml from tube 4AP-4 into tube 4AP-5, cap it, and mix by shaking.

12. Transfer 1 ml of sample from the dilution tube 4AP-5 into the inoculated molten soft agar tube 4. As in step 6, quickly cap the soft agar tube, and gently invert it a few times to mix it. Then immediately pour the contents of the tube onto the agar plate labeled 4AP-5 as described in step 6.

Team C

13. At exactly 8 minutes, quickly prepare serial dilutions (AP-2 to 8AP-5), using a fresh pipette for each of the following transfers: Transfer 1 ml of the AP-2 mixture into tube 8AP-3, cap the tube and mix by shaking; transfer 1 ml from tube 8AP-3 into tube 8AP-4, cap it, and mix by shaking; transfer 1 ml from tube 8AP-4 into tube 8AP-5, cap it, and mix by shaking.

14. Transfer 1 ml of sample from the dilution tube 8AP-5 into the inoculated molten soft agar tube X and another 1 ml of the 8AP-5 dilution into molten soft agar 5. You can use the same pipette for both transfers. As in step 6, quickly cap both soft agar tubes, and gently invert them a few times to mix them. Then immediately pour the contents of each tube onto the corresponding plates, labeled 8AP-5 and 8AP-5X, as described in step 6.

15. Once the soft agar on each plate has solidified, place them, inverted, into the 37°C incubator for 24 hours.

Third Session

Analysis of Antibody Neutralization

Examine the plates, and record the results in your laboratory report.

LABORATORY REPORT

NAME ———————————————— DATE ——————————

LAB SECTION ——————————————————————————

The Neutralization of Viruses by Antibodies

1. Describe the appearance of each of the bacteriophage plating control plates, P-5 and P-6.

2. Plate P-6 should contain identifiable plaques. Count the plaques, and complete the following table.

Plate designation	Number of plaques	Dilution factor	Volume of phage suspension plated (ml)	Phage titer (PFU/ml)

3. Given the number of plaques on the P-6 plate, approximately how many plaques (or originally, plaque-forming units) must there be on plate P-5, even though you may not be able to count them?

4. Considering the deduced number of plaques (infection events) on plate P-5, determine the extent of phage neutralization, if any, by the antiserum at the designated times. Complete the following table.

Plate designation	Antiserum-phage incubation time (minutes)	Number of plaques per plate	PFU/ml remaining infective	Percent successful phage infection	Percent phage inactivation

5. Given the data presented in the table in #4, discuss the impact that anti-T4 antibodies have on the ability of T4 phage to infect *E. coli* cells.

6. Describe the appearance of the control plate labeled 8AP-5X. What is the purpose of this control?

7. Diagram a T4 phage, and depict its attachment to and entry into its host cell. Where do you expect the anti-T4 antibodies to be binding? Draw and briefly explain your answer.

8. A teenager comes into the emergency room complaining of headache and spasms of the jaw muscles. A few weeks before, he stepped on a dirty nail, cutting his foot. Because he has never been immunized against tetanus, his physician suspects that he has a *Clostridium tetani* infection. She orders a blood culture and an injection of tetanus antitoxin, antibodies specific for the tetanus toxin. What is the purpose of tetanus antitoxin in this case?

9. Three 0.01 ml samples are taken from a liquid culture of *E. coli*. One sample, labeled A, is spread directly onto an agar plate. Another sample, labeled B, is first treated with antibodies specific for *E. coli* surface molecules, and then spread on a separate plate. The third sample, labeled C, is treated with the same antibodies and complement proteins, and is spread on a third plate. The plates are placed at 37°C overnight.

    The next day, there are about 30 colonies on plate A. Predict the results for plates B and C. Explain your answer.

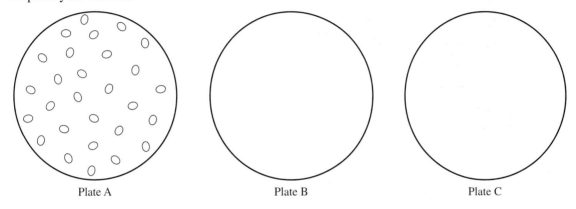

Plate A   Plate B   Plate C

10. Antibodies help dispose of foreign antigens by targeting them for destruction by opsonization or complement activation. Briefly describe each of these processes.

11. In this exercise, you used polyclonal antibodies specific for bacteriophage T4. Draw a flowchart showing how these antibodies might have been generated.

Credits

Index

Class Notes

Class Notes

Class Notes

Class Notes